'WHY DID NOT SOMEBODY TEACH ME
THE CONSTELLATIONS, AND MAKE ME AT
HOME IN THE STARRY HEAVENS?'

THOMAS CARLYLE

EX LIBRIS

....

PYXIS
OCTANS
HYDRUS
ARA
CANIS M
ANDROMEDA
APUS
TRIANGU
ARIES
AQUARIUS
VELA
RETICULUM
VULPECULA
AQUILA
LYRA
VIRGO
NOR
MUSCA
ANTLIA
AURIGA
BOÖTES
MICROSCOPIUM
MONOCEROS
CANCER
CAELUM
CRUX
ICTOR
PISCES
CEPHEUS
HYDRA

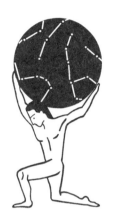

STORIES IN THE STARS
AN ATLAS OF CONSTELLATIONS

SUSANNA HISLOP
ILLUSTRATED BY HANNAH WALDRON

PENGUIN BOOKS

PENGUIN BOOKS

An imprint of Penguin Random House LLC
375 Hudson Street
New York, New York 10014
penguin.com

First published in Great Britain by Hutchinson,
a division of The Random House Group Limited, 2014
Published in Penguin Books 2015
Published by arrangement with Hutchinson

While not intended as a field guide, every effort has been taken to ensure accuracy throughout *Stories in the Stars*.

LIBRARY OF CONGRESS CATALOGING-IN-PUBLICATION DATA
Hislop, Susanna.
Stories in the stars : an atlas of constellations / Susanna Hislop ; illustrated by Hannah Waldron.
pages cm
Includes index.
ISBN 978-0-14-312813-7
1. Stars—Atlases.   2. Stars—Mythology.   3. Constellations—Atlases.
4. Constellations—Mythology.   I. Title.
QB65.H57 2015
523.802'23—dc23
2015018806

Printed in the United States of America
1   3   5   7   9   10   8   6   4   2

DESIGN AND STAR MAPS
BY WILL WEBB DESIGN

TUCANA
IS MAJOR
ERIDANUS
PUPPIS
SAGITTARIUS
LUPUS
CAMELOPARDALIS
NGULUM
PERSEU
CANIS MINOR
CARINA
TELESCOPIUM
DELPHINUS
INDUS
SAGITTA
COMA BERENICES
URSA MAJOR
NORMA
ORION
CHAMAELEON
EQUULEUS
CASSIOPEIA
CORONA BOR
CENTAURUS
CORVUS
CIRCINUS
CANES VENATICI
CAPRICORNUS
LEPUS
CRATER
LEO MINOR
LIBRA
LACERTA
SCULPTOR

CONTENTS

X I SPY IN THE NIGHT SKY

2 ANDROMEDA
AND/ANDROMEDAE,
THE CHAINED MAIDEN

6 ANTLIA
ANT/ANTLIAE, THE AIR PUMP

8 APUS
APS/APODIS, THE BIRD OF PARADISE

10 AQUARIUS
AQR/AQUARII, THE WATER-BEARER

14 AQUILA
AQL/AQUILAE, THE EAGLE

16 ARA
ARA/ARAE, THE ALTAR

18 ARIES
ARI/ARIETIS, THE RAM

20 AURIGA
AUR/AURIGAE, THE CHARIOTEER

22 BOÖTES
BOO/BOÖTIS, THE HERDSMAN

24 CAELUM
CAE/CAELI, THE SCULPTOR'S CHISEL

26 CAMELOPARDALIS
CAM/CAMELOPARDALIS, THE GIRAFFE

28 CANCER
CNC/CANCRI, THE CRAB

30 CANES VENATICI
CVN/CANUM VENATICORUM,
THE HUNTING DOGS

32 CANIS MAJOR
CMA/CANIS MAJORIS, THE GREATER DOG

34 CANIS MINOR
CMI/CANIS MINORIS, THE LESSER DOG

36 CAPRICORNUS
CAP/CAPRICORNI, THE SEA GOAT

38 CARINA
CAR/CARINAE, THE KEEL (OF ARGO NAVIS)

40 CASSIOPEIA
CAS/CASSIOPEIAE, THE ETHIOPIAN QUEEN

42 CENTAURUS
CEN/CENTAURI, THE CENTAUR

46 CEPHEUS
CEP/CEPHEI, THE ETHIOPIAN KING

48 CETUS
CET/CETI, THE SEA MONSTER (WHALE)

52 CHAMAELEON
CHA/CHAMAELEONTIS, THE CHAMELEON

54 CIRCINUS
CIR/CIRCINI, THE COMPASSES

56 COLUMBA
COL/COLUMBAE, THE DOVE

58 COMA BERENICES
COM/COMAE BERENICES, BERENICE'S HAIR

60 CORONA AUSTRALIS
CRA/CORONAE AUSTRALIS,
THE SOUTHERN CROWN

62 CORONA BOREALIS
CRB/CORONAE BOREALIS,
THE NORTHERN CROWN

64 **CORVUS**
CRV/CORVI, THE CROW

66 **CRATER**
CRT/CRATERIS, THE CUP

68 **CRUX**
CRU/CRUCIS, THE (SOUTHERN) CROSS

70 **CYGNUS**
CYG/CYGNI, THE SWAN

72 **DELPHINUS**
DEL/DELPHINI, THE DOLPHIN

74 **DORADO**
DOR/DORADUS, THE GOLDFISH

76 **DRACO**
DRA/DRACONIS, THE DRAGON

78 **EQUULEUS**
EQU/EQUULEI, THE LITTLE HORSE

80 **ERIDANUS**
ERI/ERIDANI, THE RIVER

82 **FORNAX**
FOR/FORNACIS, THE FURNACE

84 **GEMINI**
GEM/GEMINORUM, THE TWINS

86 **GRUS**
GRU/GRUIS, THE CRANE

88 **HERCULES**
HER/HERCULIS, THE KNEELING HERO

90 **HOROLOGIUM**
HOR/HOROLOGII, THE PENDULUM CLOCK

92 **HYDRA**
HYA/HYDRAE, THE WATER-SNAKE

96 **HYDRUS**
HYI/HYDRI, THE LESSER WATER-SNAKE

98 **INDUS**
IND/INDI, THE INDIAN

100 **LACERTA**
LAC/LACERTAE, THE LIZARD

102 **LEO**
LEO/LEONIS, THE LION

104 **LEO MINOR**
LMI/LEONIS MINORIS, THE LION CUB

106 **LEPUS**
LEP/LEPORIS, THE HARE

108 **LIBRA**
LIB/LIBRAE, THE SCALES

110 **LUPUS**
LUP/LUPI, THE WOLF

112 **LYNX**
LYN/LYNCIS, THE LYNX

114 **LYRA**
LYR/LYRAE, THE LYRE

116 **MENSA**
MEN/MENSAE, THE TABLE MOUNTAIN

118 **MICROSCOPIUM**
MIC/MICROSCOPII, THE MICROSCOPE

120 **MONOCEROS**
MON/MONOCEROTIS, THE UNICORN

124 **MUSCA**
MUS/MUSCAE, THE FLY

126 **NORMA**
NOR/NORMAE, THE SET SQUARE

128 **OCTANS**
OCT/OCTANTIS, THE OCTANT

130 **OPHIUCHUS & SERPENS**
OPH/OPHIUCHI, THE SERPENT BEARER
SER/SERPENTIS, THE SERPENT

134 **ORION**
ORI/ORIONIS, THE HUNTER

136 **PAVO**
PAV/PAVONIS, THE PEACOCK

138 **PEGASUS**
PEG/PEGASI, THE WINGED HORSE

142 **PERSEUS**
PER/PERSEI, THE HERO

144 **PHOENIX**
PHE/PHOENICIS, THE PHOENIX

146 **PICTOR**
PIC/PICTORIS, THE PAINTER'S EASEL

148 **PISCES**
PSC/PISCIUM, THE FISHES

152 **PISCIS AUSTRINUS**
PSA/PISCIS AUSTRINI,
THE SOUTHERN FISH

154 **PUPPIS**
PUP/PUPPIS, THE STERN
(OF ARGO NAVIS)

156 **PYXIS**
PYX/PYXIDIS, THE SHIP'S COMPASS

158 **RETICULUM**
RET/RETICULI, THE NET

160 **SAGITTA**
SGE/SAGITTAE, THE ARROW

162 **SAGITTARIUS**
SGR/SAGITTARII, THE ARCHER

164 **SCORPIUS**
SCO/SCORPII, THE SCORPION

166 **SCULPTOR**
SCL/SCULPTORIS, THE SCULPTOR

168 **SCUTUM**
SCT/SCUTI, THE SHIELD

170 **SEXTANS**
SEX/SEXTANTIS, THE SEXTANT

172 **TAURUS**
TAU/TAURI, THE BULL

174 **TELESCOPIUM**
TEL/TELESCOPII, THE TELESCOPE

176 **TRIANGULUM**
TRI/TRIANGULI, THE TRIANGLE

178 **TRIANGULUM AUSTRALE**
TRA/TRIANGULI AUSTRALIS,
THE SOUTHERN TRIANGLE

180 **TUCANA**
TUC/TUCANAE, THE TOUCAN

182 **URSA MAJOR**
UMA/URSAE MAJORIS, THE GREAT BEAR

186 **URSA MINOR**
UMI/URSAE MINORIS, THE BEAR CUB

188 **VELA**
VEL/VELORUM, THE SAIL
(OF ARGO NAVIS)

190 **VIRGO**
VIR/VIRGINIS, THE MAIDEN

194 **VOLANS**
VOL/VOLANTIS, THE FLYING FISH

196 **VULPECULA**
VUL/VULPECULAE, THE LITTLE FOX

200 **ACKNOWLEDGMENTS**
201 **INDEX**
210 **ABOUT THE AUTHOR AND ILLUSTRATOR**

I SPY IN THE NIGHT SKY

··

I AM STANDING, in the middle of a field, somewhere. I lost all my friends hours ago and this is long before the era of the ubiquitous mobile phone, let alone the possibility of a fifteen-year-old owning one. Mud seeps into my Pumas, making its way through the sagging plastic bags I have duct-taped around them in an improvisation of waterproofing.

I know the words inside out. I belt them out at a volume and intensity that I feel (alongside all the other thousands of people vibrating against one another like particles of some dense gas trapped in a bell jar, sometimes jostling into one another, sometimes walking off for a beer) is in direct proportion to my unrivaled passion for the four stars in front of me.

Although I have sung these words hundreds of times—in my bedroom, at pheromone-heavy house parties, on the unaccomplished edge of the sports pitch—I have no idea what they mean. I couldn't spell them or write them down: to me they are just sounds and syllables.

So I have no idea that Alex James is singing about astronomy. Or that the strange words in the Blur song I am wailing far out into the night sky are the names of the moons, planets, and stars above me. Even if I did know the meaning of the lyrics searing themselves into my imagination, I doubt I would look up: it is raining heavily and large globules of water are pounding down on the hood of my Adidas raincoat. But if I were to shift my gaze away from the bright lights on the stage, and up to the June night, I might see Scorpius and Sagittarius approaching their highest points, or the sparkling diadem of Corona Borealis high above me, or even Altair and Vega, two of the very stars whose names mean so much to me, even in abstract song, and which, along with Deneb, form the Summer Triangle currently rising in the darkness.

For now, though, I am singing. I am singing unfathomable words that have been passed on to me in song to explain the joy, vastness, and mystery of the universe, and they make perfect sense.

···

In about 150 AD, an Egyptian astronomer, mathematician, and geographer of Greek descent living in Alexandria, named Klaudios Ptolemaios, created an astronomical treatise unprecedented in size and scope. Not only did it represent the apex of Greco-Roman wisdom on the subject, but it shaped and determined the way that we still gaze up at the stars today. For, at its center, was a catalogue of more than a thousand stars arranged into forty-eight constellations that form the basis of the system we still use to chart the sky. Based largely on the observations of a second-century BC Greek polymath, **Hipparchus**, it was in some ways a swan song to Hellenic astronomy; by 8 AD, the home of that ancient science was no longer in Alexandria but in Baghdad. Luckily, the *Mathēmatikē Syntaxis*, as this monumental treatise was called, was preserved in Arabic manuscripts. Swiftly becoming the astronomer's bible, it came to be known by its Arabic title, the *Almagest*, while the man who made it—about whom very little is actually known—became the notorious and often mythologized **Ptolemy**.

Like history's fascination with Ptolemy himself, our curiosity about the stars is as much about story as it is about science. While any astronomical scheme carving up the firmament represents the scientific achievement of the society that created it, it also represents its culture: the way in which a people plot the stars is a distillation of their collective imagination. As the English artist and writer John Berger has written:

> Those who first invented and then named the constellations were storytellers. Tracing an imaginary line between a cluster of stars gave them an image and an identity. The stars threaded on that line were like events threaded on a narrative. Imagining the constellations did not of course change the stars, nor did it change the black emptiness that surrounds them. What it changed was the way people read the night sky.

For all the impressive empiricism of Ptolemy's great masterpiece, it is hard to imagine that it would have been preserved—and loved—for thousands of years if it wasn't as brilliant a compilation of myth as of mathematics.

It was not Ptolemy who first told the story of a wild beast growling across the night sky, or a hunter leaping through the heavens with his dogs at his heels, but it was Ptolemy who put the bear of Ursa Major and the belted figure of Orion on the map. We can never know exactly how the tales told thousands and thousands of years ago by our ancestors—looking up at the vast darkness above them, whether from deserts or mountains or the sandy streets of ancient cities—morphed into the legends that Ptolemy identified so definitively in his Almagest; nor how the animals, gods, and heroes worshipped in Assyria, Babylonia, or ancient Egypt made their slippery ways across seas and centuries and into the Greek zeitgeist. Nor can we ever fully solve the riddle of how these characters changed their names and became Roman, thus creating endless confusion about whether we should call the hero who carried out twelve fearsome labors Hercules or Heracles, or whether the queen of the gods is **Hera** or **Juno**. (I've opted for an intentionally capricious mix throughout.) Even after the ancients had swallowed up the star-lore of Mesopotamia and spat it out as their own, the Arabs, medieval monks, intrepid voyagers of the sixteenth century, and the telescoped astronomers of the Enlightenment all had a good tinker with the sidereal stories of the past. A nineteenth-century mapmaker called Julius Schiller once tried to Christianize the sky by giving all the constellations biblical names and stories; while a twentieth-century globe maker reconfigured their narrative coordinates to tell the story of Alice's Adventures in Wonderland. And all this is only within the Western view of things. The ancient Chinese had an equally complex and entirely different astronomical system, while entire universes of indigenous myths, so often ignored by colonizing forces, have only recently begun to be told outside of the cultures that invented them.

Moreover, continuing developments in astronomy and navigation down the ages meant that whole new constellations had to be invented, and their creators mapped their own consciousnesses onto the expanded celestial cartography—whether they were European explorers discovering whole new continents and honoring their exotic species or eighteenth-century scientists spotting hitherto unknown galaxies and commemorating the instruments with which they did so. All of which meant that by the beginning of the twentieth century, there was a vast—and often conflicting—array of not just stories but also star atlases, cataloguing and charting the night sky. While the mysterious layers of folklore, fact, and fiction underpinning the constellations are delightful for the storyteller, they are not always so helpful for the stargazer. Cartographers across centuries and continents varied not only in the way in which they defined or illustrated the figures in the sky but in the very names and numbers of the constellations, and the stars within them. In addition, there was widespread confusion (as there still is today) about the difference between an asterism—which is just a pattern of stars, like the Plow—and a constellation, such as Ursa Major, which is a segment of the night sky (although historically often also seen as an image) and all that is contained within it. While the Greeks and all astronomers following in their wake (but interestingly, not several non-Western cultures) defined the constellations informally by the connect-the-dot shapes of the animals, gods, and heroes they saw in the stars, by the beginning of the twentieth century, this antiquated, imprecise system couldn't keep up with the pace at which astronomers—and their increasingly sophisticated technology—were discovering new stars.

So in 1922, the International Astronomical Union (the IAU, which had been founded three years earlier) set about clearing up the confusion and decided upon the eighty-eight formalized constellations that we have today. Furthermore, they commissioned a Belgian astronomer called **Eugène Delporte** to create a definitive map of the historically contested constellation boundaries. By 1930, an official and wholly scientific way of charting the night sky had been agreed upon internationally: a constellation was no longer a pattern of stars, drawn together by imaginary lines, but an area of the celestial sphere, precisely located.

This book is not about astronomical rigor (and if it was, I would certainly be the wrong person to be writing it: it is the stories, not the science of the stars that I know). Nevertheless, illustrator Hannah Waldron, designer Will Webb, and I have incorporated some simple astronomical elements into this celestial atlas of the imagination. The dotted lines around the constellations in Hannah's illustrations represent the official IAU boundaries. Although the IAU does not officially recognize asterisms, they do suggest "traditional" links between the stars that form their patterns, and we have followed these lines for the most part. Hannah has taken the same wonderfully idiosyncratic liberty with the figures she has drawn as did the illustrators of the exquisite star atlases of the past; and she has created a pattern within the boundary of each constellation inspired by and illuminating the stories I tell. Speaking of luminosity, the orange stars making up the figures that Hannah has drawn are represented with their "apparent magnitude" as seen from Earth—the lower a star's magnitude number, the more brilliant it appears. The different sizes of the orange dots indicate their varying brightness as indicated by a scale at the bottom of each page. Stars within the IAU boundary of each constellation that are not joined in a pattern, but which are brighter than magnitude 4.0, are also represented by small blue and white circles within the dotted boundary lines and textured patterns of each illustration, but have not been allocated specific magnitudes (are you still with me?). Traditionally, Greek letters have been used to mark individual stars and to designate their brightness, α signaling a constellation's lucida (its most luminous star) followed by β, γ, etc. Although we have not labeled them as such on the illustrations, I do sometimes refer to them in the text. In which, incidentally, the names of constellations are highlighted in orange, and mythological and "real" figures occurring in more than one story in bold. Hopefully, you will have fun, as I have, connecting the dots.

Negotiating the stars into a narrative is one thing, but getting to grips with celestial cartography is not so simple; and there is one last piece of astronomy that is perhaps helpful to make sense of it. The celestial sphere is an imaginary sphere projected above the Earth, creating a sort of dome by which astronomers measure the night sky. This is divided into the northern and southern hemispheres—from which regions the view of the stars is as divergent as the cultures and myths that describe them. At any given moment, we can see roughly half of the sky—the Earth's rotation, and our particular location, determines the other half that is hidden under our feet. In other words, the daily rotation of our planet, and its annual orbit around the Sun, governs the view we have of the stars.

This simple motion is what brings us the romance of the night. The constellations shift with the seasons, so that different images are played out in the darkness at different times of year: like a magic lantern, throwing up not shadows but lights, bright, bright lights, illustrating the legends of generations.

● ● ●

I am walking to school and having an argument with my friend. She has green hair and a permanent scowl and is also a committed Blur fan. We have the same argument, often ferociously, almost every morning. Which is better: art or science? Fifteen years from now she will go on to seriously challenge my argument for the prevalence of art by spending a perilous year in Uzbekistan fighting multi-drug-resistant tuberculosis. But really, the two are inseparable. Stories are science and science is story. We look around us, we notice patterns, we try to make sense of them.

And, we look up.

This is the glory of stargazing for the storyteller: above us is a blank page in negative. A jet canvas pricked with white dots, and a rag-bag of myths, religions, lullabies, and fairy tales with which to join them up. A whole universe of stories ready to steal, which are as unstable as the stars themselves—shining and magical but soon to explode and re-form from the dust and gas of history into new stories altogether.

SUSANNA HISLOP

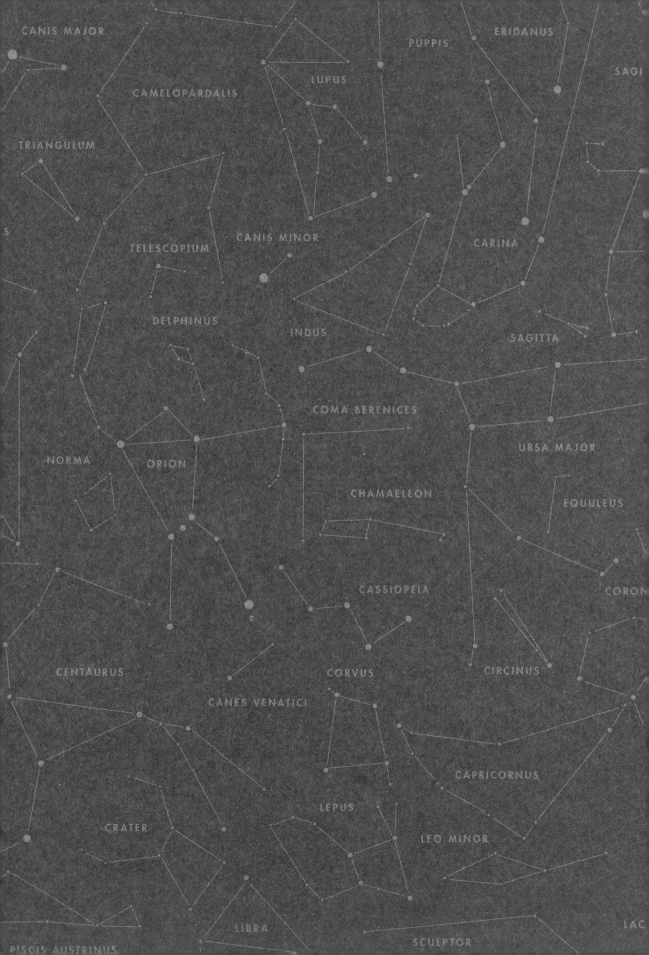

THE CONSTELLATIONS

ANDROMEDA

AND/ANDROMEDAE, THE CHAINED MAIDEN

RANK IN SIZE: 19
ASTERISMS: THE BASEBALL DIAMOND, FREDERIK'S GLORY, THE GREAT
SQUARE, THE LARGE DIPPER, THE THREE GUIDES

LOOK UP and as far away as you can see. Can you spot it? Something flickering in faraway intensity? Something 2.5 million light-years away that is hurtling toward you at 300 kilometers a second?

Squint really hard . . .

There it is. The Andromeda galaxy with its celestial motorway of a name: the "M31." It is the most distant object you can see with your naked human eye and the closest spiral galaxy to Earth.

Hurtling eternally toward *me* is Cetus, the deranged sea monster: a dragonfish, a sea serpent, a great whale, but always a she—they always are, fishy monsters, aren't they? If you look up from the northern hemisphere in late autumn at about ten o'clock—or at eight o'clock in mid-December if that's past your bedtime—you will see fierce Cetus rising toward the ecliptic, lunging toward me from the southern depths, with only the twin fish of Pisces standing in her way.

And here am I, the original maiden in distress. The Woman Chained, who awaits her knight in shining armor; whose *own parents* strung her to a *rock*. (To see how I ended up in this star-framed family horror story, read about my mother, Cassiopeia, the bitch.) The horror of my fate still lingers in the Sanskrit tongue: their word *medha* carry the bloody echoes of ancient sacrifice. But if you name a girl Medha that means she has an intellect illuminated by love. Which I like to think is about right.

Because this much is true, ladies and gentlemen: I was saved by love. Strapped to the rocks of Joppa and screaming out to sea, I caught the beady eye of Perseus. Apparently I blushed in shame and couldn't speak—but that's a lie. I blushed in raging lust.

I apologize for trampling all over the history books with the not-so-virgin truth of my girlhood—a cheeky fumble in the olive groves here, a naughty nuzzle behind the chariot shed there—but those poets get on my nerves. When gorgeous Percy swooped to my rescue on a massive white horse (not fluttering on winged sandals, **Ovid**) I was gloriously naked except for some impressive bling—something you understood when you started your star atlases, before modesty shrouded the centuries and you covered me up. Like those Arabian astronomers who, scared to draw the human form, turned me into a fat little sea calf. Thanks a bunch. They kept me in chains though, of course.

I think I like the way Rubens painted me best (sticking to the seal idea, I see, all that blubber). But if you want a sense of how voluptuously large I really am, that spiral galaxy your scientists have so romantically named M31 is nestled safely close to my right hip.

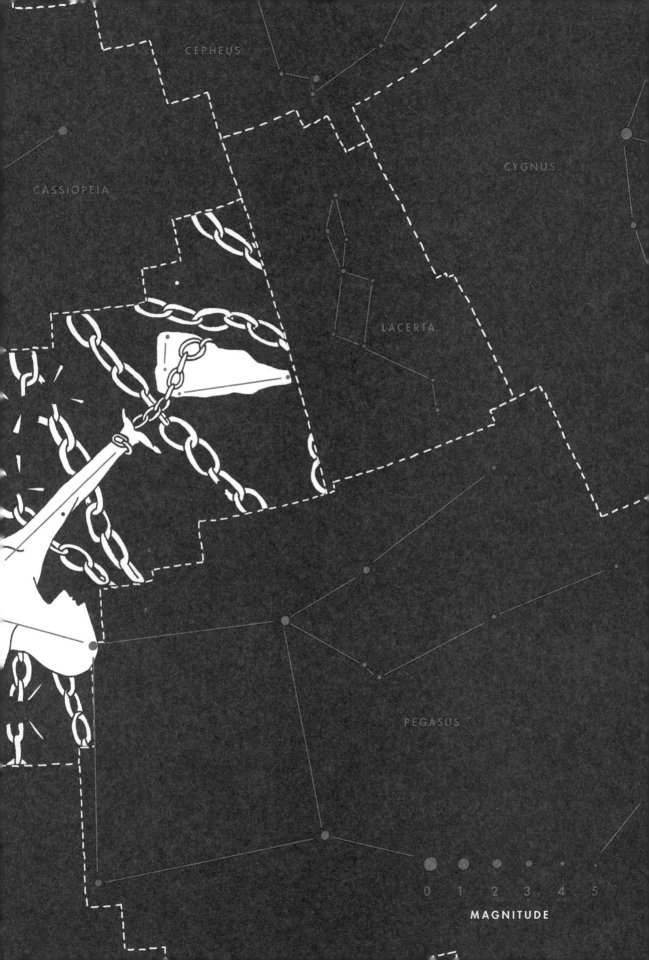

CEPHEUS

CYGNUS

CASSIOPEIA

LACERTA

PEGASUS

MAGNITUDE

0 1 2 3 4 5

ANTLIA

ANT/ANTLIAE, THE AIR PUMP

RANK IN SIZE: **62**
ASTERISMS: **NONE**

Twinkle, twinkle, little bat!
How I wonder what you're at!
The Mad Hatter, *Alice's Adventures in Wonderland*

SICK, PERHAPS, of the ancients pumping the hot air of myth into the sky, the astronomer **Nicolas Louis de Lacaille** worked indefatigably to place instead the tangible achievements of science in the firmament. He might not, therefore, be as thrilled as me to discover the imaginative coup achieved recently by inspired globe makers Greaves & Thomas. They have designed and created an exquisite celestial globe that maps the night sky using characters from *Alice's Adventures in Wonderland* and *Through the Looking-Glass*. At once witty and ingenious, what is strange is how much sense it makes of Lewis Carroll's nonsense.

For Alice is of course a young girl (Virgo) who falls down a deep, time-warping black hole ("Either the well was very deep or she fell very slowly") and is instructed in the ways of a strange universe by a whole host of anthropomorphic creatures. She meets a March Hare (Lepus) and a pair of twins (Gemini) called Tweedledum and Tweedledee; witnesses a fight between a lion (Leo) and a unicorn (Monoceros); and sees a crab (Cancer) run around in circles in the Caucus Race. The Borogove bird is an exotic bird with a large beak that would have appeared as wonderful to Alice as would the toucan (Tucana) to sixteenth-century explorers; while the tea from the Mad Hatter's teapot flows as endlessly as the water pouring from the jar in Aquarius. The Mad Hatter even asks Alice to riddle why a crow (Corvus) is like a writing desk.

When the Alice globe's maker, James D. Bissell-Thomas, first conceived of the project and imagined the Herculean task that he had ahead of him, he thought that it might be a pipe dream. But the more uncanny celestial references he discovered in *Wonderland*, the more convinced he became that the author of Alice's adventures must have had the constellations in mind. As a mathematician and Oxford don, Charles Dodgson (Carroll's real name) was well acquainted with astronomy: there were many books on the subject in his library and he even owned his own telescope.

But what of Antlia? Lacaille may be turning in his grave to see that the outline of his air pump, that triumph of physics invented by Denis Papin in the 1670s—which enables air to be sucked out of it, creating a vacuum—can just as easily be imagined as the hookah pipe smoked by Carroll's laconic Caterpillar, the crawling animal (or is it the water-snake Hydra?) so near to it in the sky.

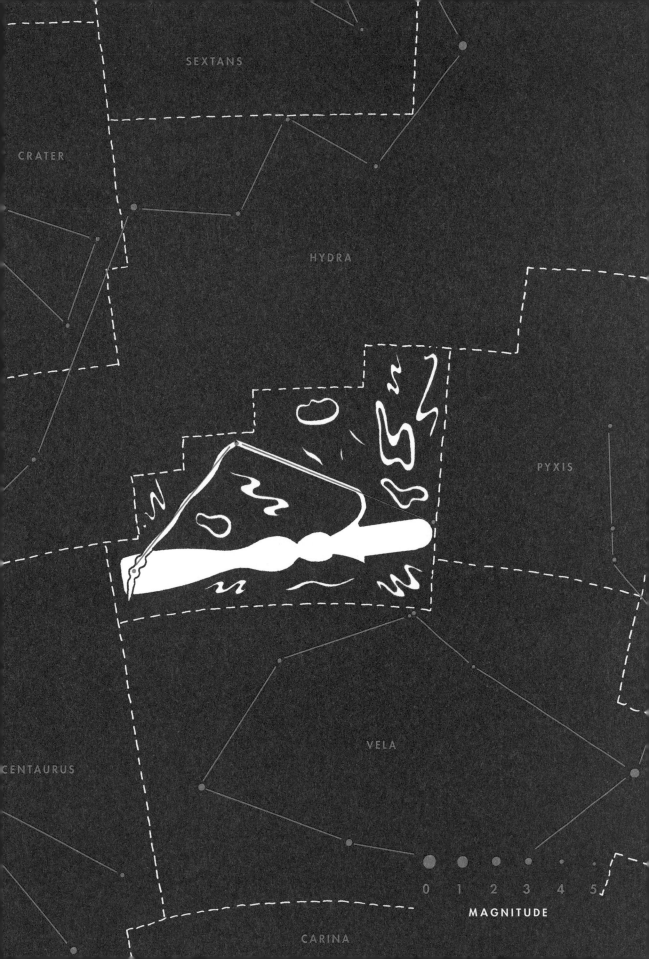

CRATER

HYDRA

PYXIS

CENTAURUS

VELA

CARINA

● ● ● ● · ·
0 1 2 3 4 5

MAGNITUDE

APUS

APS/APODIS, THE BIRD OF PARADISE

RANK IN SIZE: **67**
ASTERISMS: **NONE**

YOU ARE STANDING in the British Museum, looking at an open wooden box out of which bursts red and gold plumage. In its center a small taxidermied bird perches morbidly on a vertical branch. Imprisoned behind glass on a grizzly London afternoon, this bird of paradise feels about as far away as it is possible to feel from its native Papua New Guinea.

You gaze idly at the exhibit label. You are not particularly interested by the fact that it is on loan from the Natural History Museum, or that it was donated to that venerable institution by a Lady Lyttleton in 1931. You wonder why posh people are so keen on stuffing animals and sticking them on their walls. Lying next to the bird in his box, there are what appear to be several bodiless birds—that is, heads adjoined to feathers but missing their feet, wings, and torsos. This is the bright plumage that initially caught your eye and led you away from the graying newt you had previously been struggling to get to grips with.

What you don't realize is that it was no English gent who sliced these brilliant feathers from their corpses but a series of Oceanian tribesmen. You have no notion of the fact that in 1821 a young man in love used one of their skins to adorn his wedding outfit. Or that he spent days hunting for the perfect bird, his bride flushed with excitement when she caught sight of her handsome groom. Or that the other was sold by a village elder to feed his dying wife, before being worn in a Dutchwoman's hat in 1788 on the day she was trampled to death by a horse and carriage.

I might of course not be telling you the absolute truth. But what you can be sure of is that the people of New Guinea have been cutting off the feet and wings of birds of paradise and using their plumes in their dress and rituals for more than 2,000 years; and that European explorers have been trading these skins since **Ferdinand Magellan** first circumnavigated the Earth between 1519 and 1522. One of his crew, Antonio Pigafetta, described how "The people told us that those birds came from the terrestrial paradise, and they call them *bolon diuata*, that is to say, 'birds of God'." When these mysterious creatures made their way to the west (the name Apus comes from the Greek *apous*, meaning "footless"), astonished Europeans thought that they must spend their lives floating through the air and drinking dew; they were even briefly believed to be the mythical Phoenix.

The astronomers of ancient China also saw a peculiar bird in these southern stars: where **Pieter Dirkszoon Keyser** and **Frederick de Houtman** spied a bird of paradise, they saw a "curious sparrow."

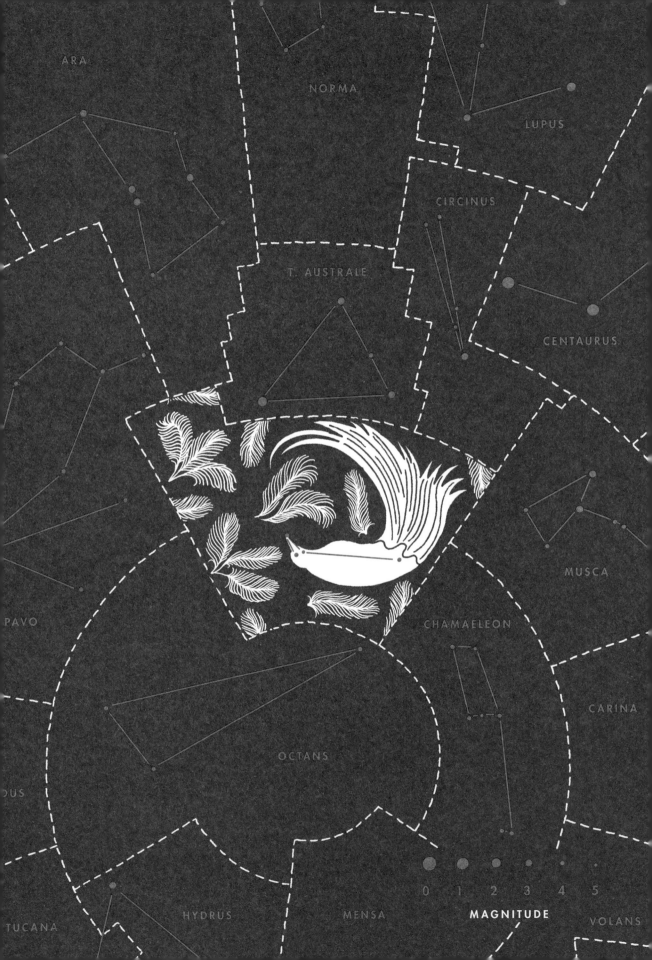

ARA

NORMA

LUPUS

CIRCINUS

T. AUSTRALE

CENTAURUS

PAVO

MUSCA

CHAMAELEON

CARINA

OCTANS

US

TUCANA

HYDRUS

MENSA

VOLANS

0 1 2 3 4 5

MAGNITUDE

AQUARIUS

AQR/AQUARII, THE WATER-BEARER

....

RANK IN SIZE: **10**

ASTERISMS: **THE WATER JAR**

....

THE STARS have an ocean just as we do. A watery kingdom high up in the heavens swimming with slippery fish and strange sea beasts. Pisces, Cetus, and Hydra lurk in its depths, and Delphinus crests its waves. And in among this sea of stars, holding sway, is Aquarius the water-bearer, the eleventh constellation of the zodiac.

As the Anglo-Saxons well knew, water has been flowing from "se Waeter-gyt" (the water-pourer) since time began. Four millennia ago, the Babylonians looked up and saw a brimming urn decanting water into the stars. For them, this constellation brought the curse of rain. The Egyptians saw it as the god of the Nile. The Hindu zodiac called their water-pitcher "kumbha," and in the rare 1703 tome, the *Meteorologiae* by "Mr. Cock, Philomathemat," Aquarius goes by the moniker "skinker."

But the Greeks, of course, have the lustiest tale. Enter leggy young **Ganymede**, the most delicious-looking mortal alive.

...

"Lo! Ganymede appears with a foaming tankard of ale."
Punch, 1841

One sunlit morning, as the sprightly youth Ganymede gamboled through the fields, flicking his blond locks and tending innocently to his sheep, **Zeus**—who had taken a liking to the comely boy—sent his eagle Aquila down to Earth to abduct the object of his affections.

Imagine Ganymede's terror, when this vicious bird (the same creature that had pecked at **Prometheus**'s innards) appeared on the horizon beating its vast wings in his direction. Little did he know, as the eagle's razor-sharp beak loomed into view, that it would pick him up without so much as bruising his plump flesh and carry him to unimaginable bliss; and so the shaking shepherd ran from its sights as fast as he could force his feet to run, stumbling over the rocky hills and pushing aside his loudly protesting sheep.

It was, of course, a futile attempt. Grasping our pretty hero in his talons, Aquila swept him up to Mount Olympus. Astonished enough that he was still alive, Ganymede was beyond all comprehension when he saw the king of the gods himself, looking straight at him with a curious look that Ganymede had never seen before—but very much liked. Taking the boy by his long-fingered, soft-skinned hand, Zeus led him through the clouds. For the next few days, or maybe years—Ganymede had no idea; time behaved very strangely up in Olympus—the god sat him by his throne and spoiled him with all the divine pleasures in his power. So overwhelmed was Zeus by love and lust that he granted Ganymede eternal youth and immortality, so that they might continue forever in this heady bliss. And what better way to spend out eternity than to sip sweet nectar dropped onto his lips by the beautiful boy? So, Zeus also gave Ganymede the privileged position of his personal cupbearer.

Unfortunately, in doing so, he ousted Hebe, the daughter of his wife, **Hera**, who until then had been entitled to that honor. Divine feathers were, as always on Olympus, ruffled: Hera was livid. Not only had her insufferable husband snubbed poor Hebe, but—and she blushed to think of it—he had seduced and ravished a *boy*. She cursed and raged with all her celestial might. Zeus realized he better not upset Hera too much but decided if he couldn't have his lover by his side, he would at least have him in the stars. So he set Ganymede, whose very name derives from the word "Rejoice!," in glory among the constellations.

So there he is in Aquarius, eternally pouring divine nectar from a golden cup, a glorious and highly popular Greek symbol of homosexual love.

PISCES

CETUS

SCULPTOR

PISCIS AUSTRI

GRUS

PEGASUS

SAGITTA

DELPHINUS

EQUULEUS

AQUILA

CAPRICORNUS

SAGITTARIUS

MICROSCOPIUM

0 1 2 3 4 5

MAGNITUDE

AQUILA
AQL/AQUILAE, THE EAGLE

RANK IN SIZE: **22**
ASTERISMS: **THE FAMILY, THE SUMMER TRIANGLE**

GOLDEN, FEROCIOUS, FAST, I am the eagle, king of the birds.

To me it is decreed to serve the Almighty **Zeus**. To fight for and to retrieve the wrathful thunderbolts of my master. To fly eternally eastward across the Milky Way.

I am the bird that snatched comely **Ganymede** in my talons and delivered him to his divine seducer. The ever-meddling poet **Ovid**—metamorphozing my tale to suit himself—has it that Zeus did the snatching; that the Almighty assumed my form to steal the young boy and flew down from heaven beating his "cunning wings." But my master did no such thing. It was me he sent, while he waited on Mount Olympus in feverish anticipation.

And it was me he sent to peck at the innards of poor **Prometheus**. A cruelty I rendered but did not condone: I am violent, I am vicious, but I dislike being unjust. I was sent to the mountains of Caucasus where that brave Titan—who had dared to steal a ball of fire from the Sun to bring light, heat, and knowledge to humankind—was chained, naked, to the rocks.

Day in, day out, I flung myself at Prometheus's flesh, piercing his very liver with my beak, hacking at his guts with my claws. Each night the immortal Titan's wounds healed—painfully, awesomely, the flesh growing back on itself in tautened folds. And each morning I returned, flying into his anguished view, to feast upon his sores: eagle-eyed, I headed straight for his abdomen. I was ferocious, I was fast, but a soldier's heart does not always obey its commands as easily as his limbs.

Finally, good Hercules took pity on the long-suffering Prometheus, and together with the wise centaur **Chiron**—who was as outraged as Hercules by Zeus's torment of this kind champion of mankind—struck a bargain with the Almighty god. Kind Chiron would give up his immortality if Zeus would set Prometheus free.

And so it was me then who took the shot for Zeus. It was me whom Hercules struck to the ground in one shot: his magic arrow driven straight into my heart.

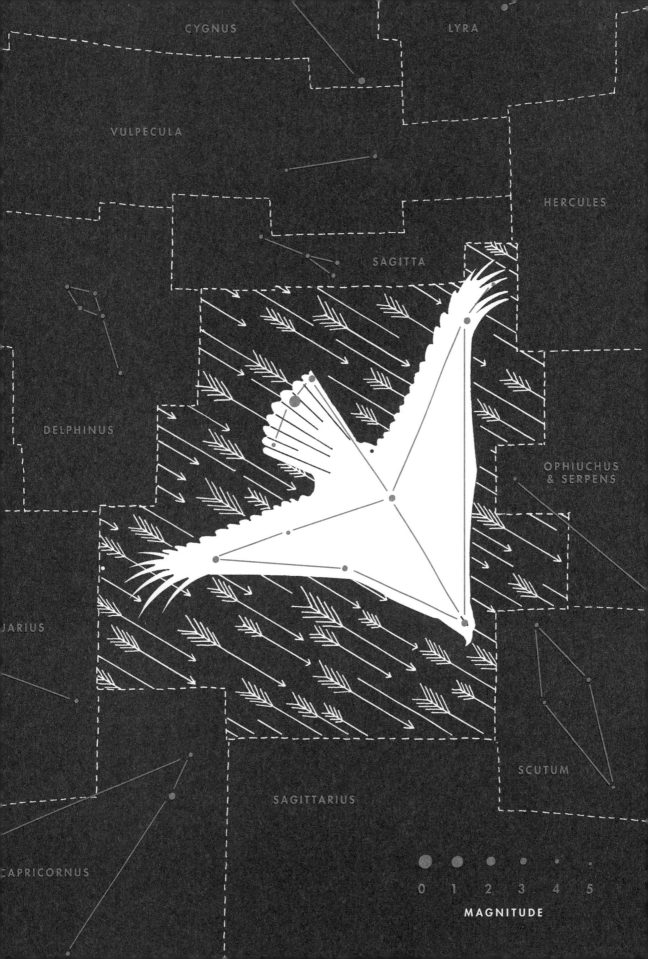

CYGNUS

LYRA

VULPECULA

HERCULES

SAGITTA

DELPHINUS

OPHIUCHUS
& SERPENS

...ARIUS

SCUTUM

SAGITTARIUS

CAPRICORNUS

● ● ● ● ● ●
0 1 2 3 4 5

MAGNITUDE

ARA

ARA/ARAE, THE ALTAR

RANK IN SIZE: **63**
ASTERISMS: **NONE**

IN THE CREAMY depths of the Milky Way lies an altar of stars. It's not that easy to spot, and it has little to distinguish itself in the way of exciting astronomical phenomena. Yet this small constellation, so seemingly insignificant that no one has even named any of its stars, has been recognized and worshipped for thousands of years. Long ago, when Iraq was part of Mesopotamia, the legend went that offerings were burned on this celestial altar to thank the gods for saving King Utnapishtim, the ancestor of the great hero **Gilgamesh**.

Utnapishtim was the wise king and priest of Shuruppak, a city on the banks of the river Euphrates. The city was as old as time and full of gods who, one day, got it into their heads to besiege the city with a flood. So Enki—the god of mischief, craft, seawater, and creation—went to Utnapishtim and told him that he must build a giant ship to escape the deluge.

Utnapishtim bent down on his knees and Enki sketched out a plan for the ship before him on the sand: it was vast, as long as it was wide and tall, and seven floors high. Utnapishtim sent for the craftsmen of Shuruppak, and they set to work as fast as their fear. Utnapishtim plied his men with ale and wine, and at sunset on the seventh day they finished the ark.

Seeing the skies beginning to turn, the king gathered together everything he could—his kith and his kin, his livestock and grain, his silver, gold, and all his loyal craftsmen—and squeezed every last being into the teeming ark. He sealed up its doors with clay, ready to set sail.

The storm hit and the land was turned to water. The ship heaved and groaned like a woman in the agonies of labor, and even the gods retreated from the flood in fear. After twelve baleful days the storm subsided, and Utnapishtim opened the hatch of the ark. He saw that it was balancing on the slopes of Mount Nimush and that his world, and all that he had known, had been swallowed by elemental rage. (A rage so fierce that the brushes of modern archaeologists have revealed its scars in the dust.) After seven days Utnapishtim sent forth a dove. When the dove returned, he sent forth a swallow; when the swallow returned, he sent a raven; and when the raven never came back, he knew that there was at last some dry land, and so he opened the doors of the ark and set all that was in it free.

In return for his piety and for saving the human race, the gods granted him immortality and a place in the heavens. And the celestial smoke that burns from Ara in his honor is the haze of the Milky Way.

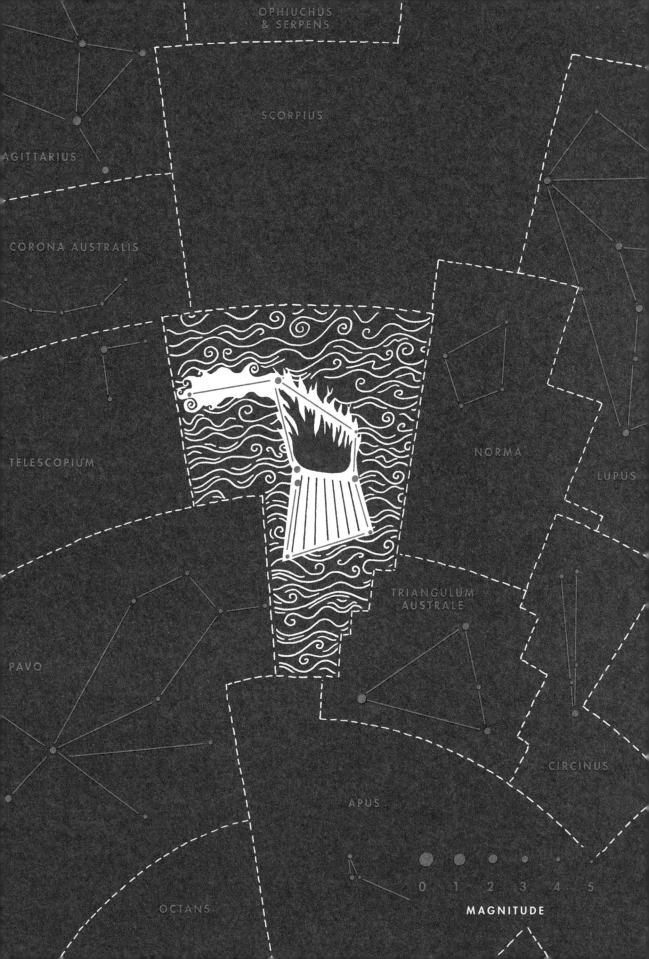

OPHIUCHUS
& SERPENS

SCORPIUS

AGITTARIUS

CORONA AUSTRALIS

TELESCOPIUM

NORMA

LUPUS

TRIANGULUM
AUSTRALE

PAVO

CIRCINUS

APUS

OCTANS

0 1 2 3 4 5

MAGNITUDE

ARIES

··

RANK IN SIZE: **39**
ASTERISMS: **THE NORTHERN FLY**

··

A IS FOR ARIES, the beginning of things. The first constellation of the zodiac and the first month of the ancient calendars. The curly-horned ram headbutting us into the year, bringing spring, light, the equinox.

It is 1664. It is the middle of the night, or rather, the very early hours of the morning, and a cantankerous Robert Hooke peers through his telescope, pursuing a recalcitrant comet. He is hungry and cold and not as rich or famous yet as he will come to be. Holding his eye to the lens, and squinting through its seventeenth-century scope, he sees something that makes his heart jump. What strange vision glitters at him through the light-years? What can he see through the dust?

Is it some commotion on Aries's brightest star—α Arietis, scientifically speaking, but named Hamal after the Arabic for "lamb" or "head of the ram" (*ras al-hamal*)? Or perhaps his telescope has focused on the stars Sheratan and Mesarthim and translated back to him not, as usual, their burning lights but instead the figures of twin horsemen—the Ashwins, divine doctors of Hindu mythology—galloping across the sky? Does Hooke spy, with his Enlightenment eye, the gods?

Perhaps he sees **Zeus**'s pet ram rescuing **Phrixus** and **Helle**, siblings of Greek mythology, from the sacrificial altar. The winged creature flies them to Colchis. Years later it will take **Jason** and a whole ship of Argonauts to bring back its Golden Fleece. Maybe the startled Hooke can even see wise **Ptolemy**, sitting in a hot room in Alexandria in AD 150, poring over his fusty tomes? He is scratching his chin while sorting a thousand stars into a catalogue of forty-eight constellations—his great work, the pinnacle of ancient astronomical wisdom, his *Almagest*.

No. It is hard to imagine Hooke, that scrupulous man of science, seeing all that. It is hard to imagine this of a man who gave his name to an incontrovertible law of physics (thereby forcing generations of GCSE physicians to hang weights from metal springs, to stare sullenly at their coils, furled like a ram's horn); it is hard to imagine the rational Robert seeing all those stories in the sky. But what he does see is no less miraculous to him. For what his telescope lands upon inadvertently while following the infuriating comet is something hitherto unknown in his universe. γ Arietis, Mesarthim—the "First Star of Aries"—is not in fact one star, but two. It is a double star. And Hooke is one of the very first people ever to have discovered this phenomenon.

He has seen a handsome twin drive his chariot across the dawn after all.

ANDROMEDA

PERSEUS

TRIANGULUM

TAURUS

PISCES

CETUS

ERIDANUS

0 1 2 3 4 5

MAGNITUDE

AURIGA
AUR/AURIGAE, THE CHARIOTEER

RANK IN SIZE: 21
ASTERISMS: THE HEAVENLY G, THE KIDS, THE WINTER OCTAGON,
THE WINTER OVAL

SOME SAY the charioteer is Erichthonius, son of Mother Earth and the fire god **Hephaestus**, who, skilled by **Athena**, was the first man to tame horses and harness them to a chariot. Some say he is Myrtilus, the crafty charioteer who met his death in an ill-fated plot to spend the night with his boss King Oenamaus's daughter.

But no one knows why he's holding Amaltheia the goat.

THE STORY OF THE MYSTERIOUS GOAT

Cronos, first of the **Titans**, was the son of **Gaia** the earth and **Uranus** the sky, and **Rhea**, his sister, was also his wife. And their child, who grew up to be the god of gods, was **Zeus**.

Now Cronos was a hungry, nasty kind of a father who devoured all his progeny as soon as they were born. So when Rhea gave birth to the tiny Zeus, and Cronos held out his hands to lift the newborn babe to his mouth, she placed, swaddled, into his greedy hands a stone instead of the infant.

Before the Titan king could see what she had done, Rhea had the baby hidden deep in a cave on Mount Ida, high up on the island of Crete.

It was not long that the little baby lay there, struggling out hungry, red-faced wails, for along came the she-goat Amaltheia, who Rhea had found as a nurse.

By name by nature, the old saying goes, and so with Amaltheia, the "tender goddess." She suckled and cared for the child Zeus until he grew up to be a strong and fine boy. So fine and strong indeed that he often forgot his own strength: one day when he was playing hide-and-seek with his gentle nurse, he accidentally snapped off one of her horns. And that became the horn of plenty we know as Cornucopia, forever spilling out food and riches as nourishing and plentiful as the milk Amaltheia gave to Zeus.

LYNX

PERSEUS

GEMINI

TAURUS

ORION

0 1 2 3 4 5

MAGNITUDE

BOÖTES
BOO/BOÖTIS, THE HERDSMAN

....

RANK IN SIZE: **13**

ASTERISMS: **THE DIAMOND (OF VIRGO), THE ICE-CREAM CONE, THE KITE, THE SPRING TRIANGLE, THE TRAPEZOID**

....

SOME SAY he's a bear driver with a club, chasing Ursa Major and Ursa Minor around the North Pole; some say he's a herdsman with a sickle and a shepherd's crook; and some say he's the man who invented the Plow. Still others argue whether his name comes from the Greek for "ox driver," or the Greek for "noisy"—as a shepherd's calls would indeed have been. Either way, the giant figure of Boötes is usually a man herding animals across the sky. In Sophocles's *Oedipus Rex*, a shepherd recalls grazing his flocks "from spring to the rising of Arcturus in the fall," and according to Hyginus's *Poetic Astronomy*, it is the shepherd **Icarius** who is remembered in this constellation.

You will remember this gregarious grape grower, who was taught to ferment his crop by the god of wine himself, from the legend of Canis Minor. We left the wild party he threw to celebrate this joyful, alcoholic discovery, just as the bright sun had risen too high for his groaning guests and their hangovers. Stumbling out of their makeshift beds on Icarius's *triklinai* (reclining couches, of the sort the Greeks liked to dine upon), his retching revelers decided their host must have poisoned them. Why else would their heads feel as if they were being hammered-and-tonged by the great fire god **Hephaestus** himself? So they murdered the innocent shepherd in his sleep. Icarius's faithful dog **Maera** whined and howled on its chain so loudly that his daughter Erigone went to unleash it. Grabbing her tunic in its teeth, the dog led her to the ditch where they had dumped her father. Devastated, she hanged herself on a nearby tree, while loyal Maera lay down to die beside its master. **Zeus** placed them in the stars as Boötes, Virgo, and Canis Minor.

The earliest classical reference to this constellation comes, however, from Homer, who recounts how Odysseus navigates away from Calypso's island using its stars. And it was not just the Greeks who sailed the seas by the light of Arcturus, Boötes's lucida and the fourth-brightest star in the sky. Western explorers, who had arrived in Pacific waters with the aid of complex astronomical instruments, were flabbergasted to learn that the ancient Polynesian "star-compass" was not, like their own sextant or astrolabe, something they could hold in their hands, but something that they kept in their heads. Directing their canoes by the rising and setting of the stars, the Polynesians knew they were at Hawaii when they saw the orange light of *Hokule'a* (their name for Arcturus, meaning "Star of Gladness") straight overhead; as they headed south, it would gradually fall from its zenith. When they found themselves paddling directly under the white light of Sirius, they knew they had reached Tahiti.

DRACO

URSA MAJOR

RCULES

CANES VENATICI

RONA BOREALIS

COMA
BERENICES

SERPENS

● ● ● ● ● · ·
0 1 2 3 4 5

MAGNITUDE

VIRGO

CAELUM

CAE/CAELI, THE SCULPTOR'S CHISEL

RANK IN SIZE: **81**
ASTERISMS: **NONE**

EVERYBODY SAID that Linda had great legs. Always had. And if her toenails were yellowing at an alarming pace, or if the flesh above her knees made her think, on occasion, of peeling the skin off a raw chicken, there was nothing that a dash of Chanel Le Vernis nail polish or a new pair of support tights couldn't do to allay her fears.

When she stepped into the shower and caught a glimpse of herself in the bathroom mirror, she saw the curve of her breasts very much as an upward angle; and if she turned her head over her shoulder as she pushed the shower door shut, she saw nothing more worrying than two gently plump buttocks through the glass.

But, she knew, it was her face that gave her away. People kept asking her if something was wrong. She had heard colleagues around the photocopier muttering something about "a permanent frown"; construction workers kept chirping at her to "cheer up, love"; and a stranger on the bus had scrunched up her face in a patronizing gesture of support.

Walking into the clinic, she almost felt her resolve sink into the thick carpet with her high-heeled feet. The woman—girl—behind the desk was ever so pretty, though, stylish, really stylish, and so well dressed. While something about the artfully arranged lilies sucking up water from an asymmetrical vase and the gentle hum of the water cooler kept her from leaving the room. If she didn't examine the fine print too anxiously, the list of risks was almost negligible, and she could choose from several neatly named packages: all at 0 percent finance!

It was unfortunate that Dr. Sinclair had had a long day. And the radio can be very distracting. It was only when he sewed the last stitches behind Linda's ears that he realized there was now a nipple where there should be a nose. And only when the suction pump vacuuming up Linda's excess fat started to splutter and splurge that he noticed her left leg was now no more than a stump, rounded off to perfection by the smooth curve of her plump left breast.

Obviously this modern morality tale of metamorphosis has little to do with astronomy: I made it up. But since **Lacaille** named this constellation after a sculptor's chisel—imbued with no more mythical status than poor, miserable Linda's big toe—I think I can be allowed a bit of artistic license.

ERIDANUS

LEPUS

FORNAX

COLUMBA

PUPPIS

PICTOR

HOROLOGIUM

RETICULUM

NA

● ● ● ● ● ●
0 1 2 3 4 5

MAGNITUDE

DORADO

CAMELOPARDALIS

CAM/CAMELOPARDALIS, THE GIRAFFE

RANK IN SIZE: **18**
ASTERISMS: **NONE**

DO GIRAFFES have flat feet? Certainly some of the long-necked creatures painstakingly illustrated on the globes and star maps of the seventeenth century are strange-looking beasts. It is hard to know why else flat-footed Peter—that is, the cartographer, astronomer, and Calvinist minister **Petrus Plancius**—decided in 1612 to put one in the sky. Petrus Plancius is not of course the name his Dutch mother gave him, as she pushed him out into the small town of Dranouter in 1552. It is the Latinized and appropriately scholarly version of that name—Pieter Platevoet (literally "flatfoot")—that he coined for himself.

Some years later, in 1624, the German astronomer Jacob Bartsch mistook this stellar giraffe for a camel. (He clearly wasn't as good at Latin as our friend Peter, who knew full well that Camelopardalis is the romanization of the Greek and means "a camel-like animal with leopard-like spots"). Bartsch surmised that it commemorated the biblical camel that carried Rebekah to her husband, Isaac, in the Book of Genesis. Now *that* is a curious tale . . .

The aging Abraham sends one of his servants to his birthplace to find a suitable wife for his son, Isaac, who is thirty-seven—very much on the shelf in those days. The servant sets off with ten camels, saddlebagged with gifts, and reaches the city as evening sets in: just at the very time—as fate would have it—that women go to draw water from the well outside the city's walls. He prays to God for a way to be sure he has found the right woman for his master and decides upon this: the woman whom he asks to let down her bucket so that he may drink, and who replies that she will sate his thirst, and who will *also* water his ten camels, will be the chosen one. Luckily, a startlingly beautiful young virgin called Rebekah appears and does just that. She runs to the well and fills her bucket and takes the water to the servant, and then she runs back to the well and fills her bucket and takes the water to the first camel, but the camel is still thirsty—because a camel can travel up to a hundred miles in a day and last up to seven days without any water, so they need gallons and gallons to fill up their humps. She runs back to the well and fills up her bucket again and again until all ten camels are full. The servant gives her a gold nose ring weighing half a shekel and two bracelets for her arms, and they go back to her family and wrap the whole thing up, and that's how Isaac gets such a wonderful wife.

URSA MINOR

DRACO

URSA MAJOR

CEPHEUS

LYNX

CASSIOPEIA

AURIGA

PERSEUS

0 1 2 3 4 5

MAGNITUDE

CANCER
CNC/CANCRI, THE CRAB

RANK IN SIZE: 31
ASTERISMS: THE ASSES, THE MANGER

UNDER THE VAST darkness of an African night sky, a dung beetle is rolling a ball of dung across the savannah. This teeny-brained scarab has sniffed out the most nutritious feces it could find, packed it into a sphere about forty times its own size, and is now transporting it back to its nest in a startlingly straight line. A friend of his once tried to roll dung in the daytime and was sizzled to death by the heat of the Sun when he'd barely set out; another was ambushed by a rival gang, who had seen the unsuspicious creature leaving the swarming dungheap, followed him, and stolen his precious cargo. So this wise scarab is doing it at night. He is journeying through the darkness, as his family has done for centuries, in a way that it took those big-brained human scientists right up until 2013 to understand: he is navigating by the stars. Using his spindly back legs to push it along, he is orientating his ball of dung in a long, straight path by the light of the Milky Way.

Does he know that 2,000 years ago ancient Egyptians were fashioning amulets in his likeness and placing them as sacred objects in their royal tombs? Does this humble dung beetle know that there was a time when his forebears were worshipped on a par with the sun god Khephri, who rolled his own precious globe—the Sun—across the sky? Does he know that he was seen as a symbol of creation and rebirth? Or that Khephri himself was often depicted as having the very same head as this sacred scarab, whom the Egyptians also honored in the stars as the constellation Scarabaeus?

Meanwhile, as this intrepid dung beetle toils across the dust of the savannah, a ragged crab is scuttling across the sand of the ocean floor. Perhaps she knows why today we call this constellation Cancer and not Scarabaeus. Perhaps her grandparents have told her the story again and again: that many moons ago, when life was better and the sea was still an Eden, free from diapers and plastic bags, one of her crustacean ancestors, Karkinos, nipped the great Herakles on the toe as he battled with the many-headed Hydra. Although this intrepid crab was swiftly crushed to death under the hero's foot, he was placed forever in the stars by Herakles's enemy **Hera**, in thanks for his small part in her divine battle.

You probably feel quite certain that neither of these creatures knows its part in the stories of the sky. But since we humans once thought that it was only seals, birds, and—of course—ourselves who navigated by the stars, I wouldn't be quite so sure.

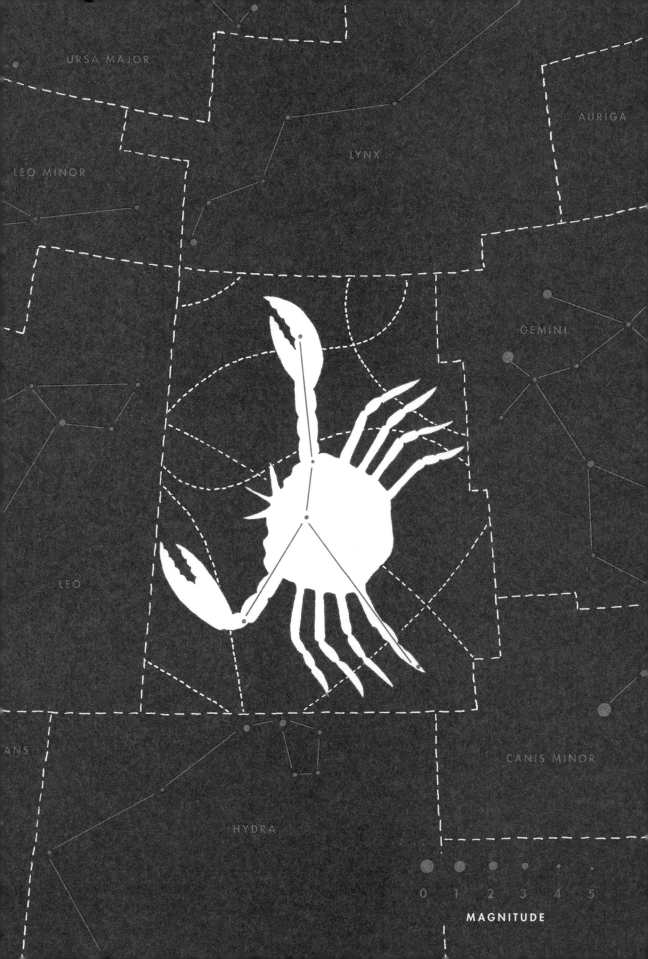

URSA MAJOR

AURIGA

LYNX

LEO MINOR

GEMINI

LEO

ANS

CANIS MINOR

HYDRA

0 1 2 3 4 5

MAGNITUDE

CANES VENATICI

CVN/CANUM VENATICORUM, THE HUNTING DOGS

RANK IN SIZE: **38**
ASTERISMS: **THE DIAMOND (OF VIRGO)**

HENRIETTA had been sitting for some time now. Naturally he had been unable to get the spaniel to stay still for very long, but that would not be a problem—they all looked the same, these inbred lapdogs. The be-studded dress, the bosom, the creamy white sloping shoulders of its mistress—that was all easy, and in any case, it was an unwritten rule—nay, *command*—that he was at liberty when it came to the exact proportions of all these. It was the princess who was the difficulty. Something in those fishbowl eyes that was impossible to get right: an intermingling of sadness, haughtiness, and flirtation that fixed itself inadequately into one of these as soon as you tried to harness it in paint.

But what a life she had led, this Minette (for that was what her affectionate brother called her in his letters). Forced to flee England at the age of three; her father executed and her mother impoverished; brought up as Roman Catholic, although the girl had been baptized in Exeter Cathedral for heaven's sake; and then finally bundled off into an unhappy marriage with a lascivious Frenchman. He wondered how well she had known her father, the old King Charles. Apparently a star shone unusually brightly that night in 1660 when Henrietta's brother returned to London and the monarchy was restored. (He remembered it well—that sense of possibility, as if the world was liberated from a puritanical palette and that he might paint in bright colors again.) The new King Charles's physician had given that star the name Cor Caroli, "Charles's Heart," in honor of their beheaded father.

The duchess gave a languid sigh. He picked up his paintbrush and carried on.

•••

What the historically contested painter of the portrait that now hangs in London's National Portrait Gallery may or may not have known was that not so many years later, in 1687, the Polish astronomer **Johannes Hevelius** was busy incorporating that same star (α Canum Venaticorum) into a new constellation. Hevelius depicted Canes Venatici as the two hunting dogs of the bear driver Boötes, snapping at the heels of Ursa Major. Curiously, the King Charles spaniel (so named because it was Charles II's favorite breed) was originally a hunting dog as well as a lapdog. That is of course before the king's choice canine—the Duke of Marlborough was also partial to a hunting pack of red and white ones—was interbred with a pug and became a flat-nosed toy dog with protruding eyes. Which makes me think of the other famous hunting dogs in the sky: Canis Major, whose brilliant star Sirius flickers red and white, and little Canis Minor, the Lesser Dog, a playful pup who runs around after its greater companion.

DRACO

URSA MAJOR

BOÖTES

COMA BERENICES

0 1 2 3 4 5

MAGNITUDE

CANIS MAJOR

CMA/CANIS MAJORIS, THE GREATER DOG

RANK IN SIZE: **43**
ASTERISMS: **THE HEAVENLY G, THE WINTER OCTAGON, THE WINTER OVAL, THE WINTER TRIANGLE**

THIS IS THE BIG DOG. The leader of the pack, with a plethora of tall tales to suit his stature. The proud hound that the ancient Mesopotamians saw tearing across the heavens in pursuit of a terrified hare—Lepus—shivering under a giant huntsman's feet. To the Greeks this huntsman was Orion, and his dog Canis Major was usually paired together with a playful companion Canis Minor. Sometimes one or the other of these two celestial canines was seen as **Maera**, the loyal dog of the brutally murdered **Icarius**. Death also haunts another Hellenic legend: that of **Cerberus**, the three-headed hellhound who guarded Tartarus, the torturous abyss in the depths of Hades. A manifestation of the ancient Egyptian god Anubis, this jackal-headed deity was the guide of the dead, weighing their souls on the scales of justice to decide their fate in the afterlife. They saw this lord of embalming and funereal rites in one specific star: the constellation's illustrious lucida, Sirius.

Sirius (α Canis Majoris) is the "Dog Star." Named for its searing, scorching light, it is the brightest star in our sky. Outshone only by a planet, it can be found in the earliest astronomical records. Known in Egypt as *Sopdet*, its heliacal rising (when, after a time of absence, it first becomes visible just before sunrise) marked the beginning of their Sothic year, just before the annual flooding of the Nile and the summer solstice, when the Sun was at its very hottest. The Greeks and the Romans absorbed this idea of a period of stifling dog days, when, as **Virgil** wrote, "the sweltering Dog Star splits the cracked fields with thirst." Sirius's heat was imagined to combine with that of the Sun, emitting strange, malign emanations that would send men "starstruck" and dogs berserk and foaming at the mouth.

The ancient Chinese also saw a rabid dog in Sirius. T'ien-lang, the "Celestial Jackal," was a wolf of the east who ravaged farmyards and had to be hunted down by the Celestial Hunting Dog, T'ien-kaou, which was also seen in the stars of Canis Major. At the same time, Alaskan Inuits called Sirius "Moon Dog"; the Seri Indians of Mexico and the Tohono O'odham of Arizona saw it as a dog chasing mountain sheep; while Inuits from the Coppermine area in the Northwest Territories saw its scintillating flame as a red fox and a white fox fighting each other. Alfred, Lord Tennyson, too, saw a skirmish in its lights: "the fiery Sirius alters hue / And bickers into red and emerald."

In 1862 the American telescope builder Alvan G. Clark discovered that Sirius is in fact a binary star and that it does indeed have a faint but faithful companion: a tiny white dwarf known affectionately as the "Pup."

CANIS MINOR

ORION

MONOCEROS

PUPPIS

LEPUS

COLUMBA

0 1 2 3 4 5

MAGNITUDE

CANIS MINOR
CMI/CANIS MINORIS, THE LESSER DOG

RANK IN SIZE: **71**
ASTERISMS: **THE HEAVENLY G, THE WINTER OCTAGON,
THE WINTER OVAL, THE WINTER TRIANGLE**

TWO BRUTAL, Bloody, and Cautionary Tales about the Lesser Hunting Dog of Orion

THE IGLOOLIK LEGEND OF SIKULIAQSIUJUITTUQ

There once was a man called Sikuliaqsiujuittuq who had more flab in his belly than all the seal blubber in the cold, dark sea. Because he was so very fat, he found it hard to get a wife, and so he had to marry his sister, who was gloriously fat. They must have had a big igloo.

While all the other men went out to hunt, Sikuliaqsiujuittuq stayed at home, terrified that he would fall through the ice. As you can imagine, everyone in the camp soon became more than a little annoyed with fat Sikuliaqsiujuittuq. Finally, when the ice got as thick as it gets, they persuaded him to go seal hunting. Happily, the frozen sea did not crack beneath his caribou-and-sheepskin-clad feet, and after a long day he prepared himself to camp out on the ice with the others. Never having done this, the naive hunter inquired as to the best way to sleep. His cunning companions answered that on a person's first night on the ice, they should have their hands tied behind their back with harpoon thong. Unsuspecting, he took their advice. As he snored away contentedly, the hunters stabbed him in his sleep. Whereupon he ascended to the sky as the bloodred star we call Sikuliaqsiujuittuq, but that is known elsewhere as Procyon, the brightest star in Canis Minor and the eighth-most brilliant (and fattest) in the sky.

THE ATTICAN LEGEND OF ICARIUS

There once was a man called **Icarius** who grew very juicy grapes. **Bacchus**, wandering about one afternoon in disguise (as the gods often did), asked if he could see Icarius's fine vineyards. So impressed was the merry deity that he decided to teach this gifted mortal how to make wine. Imagine Icarius's joy when he first tasted that heavenly liquor, when he first felt those exhilarating bubbles of mirth rise up inside him, that slow deepening into mellow ecstasy. Being a generous man, Icarius wanted to share this newfound pleasure and so held a festival for all the villagers and shepherds. Bacchus, disguised as a randy goatherd, had a wild time and—as is always the sign of a truly great party—the guests passed out all over the place. To find out the gruesome end that befell this innocent grape grower when the revelers awoke the next morning, and how his faithful dog **Maera** ended up in the sky as Canis Minor, you'll have to look to the stars of Boötes.

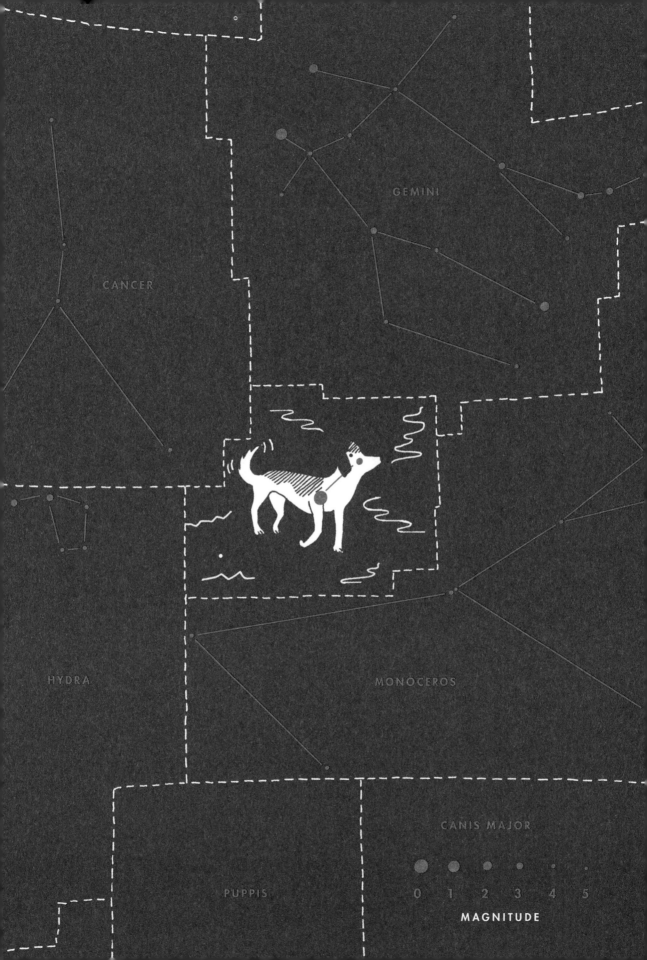

CANCER

GEMINI

HYDRA

MONOCEROS

CANIS MAJOR

PUPPIS

MAGNITUDE
0 1 2 3 4 5

CAPRICORNUS

CAP/CAPRICORNI, THE SEA GOAT

RANK IN SIZE: **40**
ASTERISMS: **NONE**

I FIND IT unfortunate that the first thing that comes to mind when someone mentions pan pipes is the theme song to *Titanic* being whistled through an unimpressed crowd of tourists in a European city and not the lyrical escapades of the playful goaty-fish god **Pan**. Similarly, I'm not sure what **Zeus**—whose life was saved from the clutches of the sea monster **Typhon** by that same sylvan deity—would make of modernity's translation of those once sweet sounds into the language of the self-help relaxation CD.

As antiquity's god of all things pastoral, Pan was usually depicted as a horned and lusty satyr with a goat's legs and cloven hooves. But medieval manuscripts portraying Capricornus with the tail of a fish as well as the head of a goat reveal this constellation's more ancient origins. There was an Assyrian-Babylonian god of wisdom called Oannes, who was half-fish half-man, while Indian astronomers of old saw both a crocodile and a hippopotamus with a goat's head in these stars. The Romans knew Capricorn as Neptuni Proles—the progeny of **Neptune** (the Roman equivalent of the Greek god of the sea, **Poseidon**)—and the constellation's lucida, Deneb Algedi (δ Capricorni), meaning the tail of the goat is only a few degrees east from the position at which the French astronomer Urbain Le Verrier calculated the planet of the same name to be.

But back to the pan pipes. Next time you find your spirits sinking in a jam-packed tourist trap, try to filter out the canned sounds of unpersuasively rustic charm with the power of myth, and imagine yourself instead by a quaint and babbling brook, dipping your toes in the stream's current in delight. Cupping your hands and lifting the icy water to your lips, you catch sight of something moving in the verdant shadows. Hiding behind a rock, you see a peculiar beast. Bearded, laughing, and clamorous, he is chasing a naked nymph. The nymph, whose name is Syrinx, is fleeing desperately from his advances, and following her escape to the edge of the trees, you see her reach the shore of a lake. The goat man catches up with the terrified dryad and lunges in her direction. Just as he reaches out to embrace her, she disappears from his grasp, and he finds himself clutching a handful of reeds. (Some sympathetic water nymphs have heard her cries.) Slumping to the ground in disappointment, he lets out a melancholy sigh. As his breath whistles through the reeds, it sounds the most beautiful and bittersweet of notes. Plucking the blades around him and cutting them to different lengths, Pan invents the instrument so subsequently debased, but to him, it's as beautiful as the nymph Syrinx he named it after.

PEGASUS

DELPHINUS

EQUULEUS

AQUARIUS

AQUILA

PISCIS AUSTRINUS

SAGITTARIUS

MICROSCOPIUM

GRUS

INDUS

0 1 2 3 4 5

MAGNITUDE

CARINA

CAR/CARINAE, THE KEEL (OF ARGO NAVIS)

RANK IN SIZE: **34**
ASTERISMS: **THE FALSE CROSS**

JASON AND THE ARGONAUTS: AN EPIC TALE
*(in which fifty Hellenic heroes set sail for Colchis in a fifty-oared ship to bring
the Golden Fleece and the ghost of* **Phrixus** *back to Greece)*
TOLD BACKWARD IN THREE PARTS
(much as **Ptolemy**'s *original constellation Argo Navis, which honored this legendary ship,
was split into three parts—*Carina, Puppis, *and* Vela—*by the French astronomer* **Lacaille** *in 1756)*

PART I: CARINA

OLD AND RAGGED, Jason sits by the rotting ship *Argo* on the shores of Colchis and contemplates his past. Once, he had stood on this same sand, strong and brave. Like all heroes, he had been handsome. Brutal, despairing, and hotheaded like all of them too. He thinks of the rowing contest they once had on the blackening hull beside him. Heracles, **Polydeuces**, **Castor**, and him—working their muscles to pride-seeking pain. Jason chuckles to himself. As if in retribution for even the memory of youthful pleasure, a sharp pain contracts in his chest.

He thinks of **Medea**. He is at last contrite. It was, he acknowledges, looking at the sea's surface darkening under patches of cloud, his fault. His children's savage murder. A stubbornness and fury in his blood he couldn't stanch. He looks absently at a gull scanning the water for prey. Who is he justifying himself to? The very deities whose names he took in vain, and under whose judgment he now wanders from place to place? He broke his promises: he had sworn to be faithful to Medea by all the gods of Olympus. He hadn't even vouched it out of love—bewitching though the sorceress was—but to use her wiles to claim the Golden Fleece. He should have known then, as she wafted her potions and hypnotized wild beasts, that she had magic in her veins but violence in her heart. Like him.

The gull dives into the waves. Medea. She had fought for him, killed for him, gained him his throne at Corinth. She had borne him seven daughters and seven sons. And he had divorced her—for a mere girl. Little wonder she had sent his naive bride a poisoned robe as a wedding gift. It was his fault, then, that when young Glauce put on the deadly dress, not just she but her father, King Creon, and all their guests erupted in unquenchable flames, while he escaped from a window in the palace. It was his fault that in retribution, the enraged Corinthians seized all his children and stoned them bloodily to death.

Climbing onto the wreck of his former glory, he prepares to hang himself on its prow. A rotting beam creaks its very last breath and keels over, hard and fatal, on Jason's head.

COLUMBA

PICTOR

PUPPIS

DORADO

VELA

VOLANS

MENSA

CHAMAELEON

TAURUS

OCTANS

MUSCA

CRUX

MAGNITUDE

0 1 2 3 4 5

CASSIOPEIA
CAS/CASSIOPEIAE, THE ETHIOPIAN QUEEN

RANK IN SIZE: **25**
ASTERISMS: **THE THREE GUIDES**

A LEGEND IN LIMPING LIMERICKS

There once was an Ethiop queen
Who was grossly concerned with her mien.
"My beauty," she swaggered,
"Is so far from haggard,
It exceeds any sea nymph's I've seen."

This so angered the sea god **Poseidon**
He struck the dark waves with his trident.
Thus emerged from the deep
A sea beast who reaped
Havoc on the whole Ethiop island.

To assuage the foul monster's curse
Cassiopeia did something much worse.
Thus she spake: "Horrid Cetus
So you no more hate us
My daughter to you I imburse."

Thus Andromeda she chained to a rock—
The poor girl in absolute shock—
But a valiant hero
Appeared and said, "Dear, oh
This story must not run amok."

Perseus rescued the beautiful virgin,
While Cassiopeia—to punish her sin—
Was placed in the stars
Her feet o'er her ****
Celestially ever to tailspin.

Yet not a hubristic queen
In a posture mildly obscene,
But rather, a lampshade,
The Inuits have splayed
Across this celestial scene.

DRACO

CAMELOPARDALIS

CEPHEUS

PERSEUS

ANDROMEDA

NGULUM

PISCES

0 1 2 3 4 5

MAGNITUDE

CENTAURUS
CEN/CENTAURI, THE CENTAUR

RANK IN SIZE: **9**
ASTERISMS: **THE SOUTHERN POINTERS**

CLEARING THROUGH the boxes in his attic, **Asclepius** found his old school report. The dusty pages took him straight back to **Chiron**'s cave on Mount Pelion. Kind Chiron, who had taught him all he knew. The terrible irony of it all: that the very person who had schooled him in the art of medicine, the tutor without whose unique wisdom he would have been nothing— let alone the great healer he had become today—should have been unable to save himself from a poisoned arrow. He flinched as he thought of the gentle centaur pulling its frothing iron from his knee.

> Although struggling a little with archery and hunting, Asclepius excels in our lessons on herbs and magic. Stiff competition from my other pupils, past and present, in particular **Jason**, **Achilles**, and Hercules, sometimes rattles his confidence, but he is progressing steadily.

Asclepius frowned. He thought back to his boyhood self. Perhaps he had been shy in the face of the legions of heroes the centaur had taught. Shouldn't he have been chastised for this weakness? But Chiron had always found it hard to be strict. Unlike the rest of his tribe, who were mostly savage and usually drunk, he was gentle and benevolent. Perhaps this was what made him such a good teacher, so trusted by the gods. Asclepius thought about his own children and wondered if he was going about things the wrong way. Hygieia, his eldest, was always so well behaved, so clean and tidy, he'd never had to worry about her (her grandfather **Apollo** had already singled her out to become the goddess of sanitation); the others occasionally gave him a bit more grief, but they muddled along. He was too harsh on his dwarf son Telesphorus, though; he knew it. To him, he was often unkind.

Rummaging through the box again he found one of his old textbooks. *Music & Harmony of the Spheres: Level One.* He leafed through the pages, appalled by the idiocy of his puerile doodles. There were several cartoonlike drawings of his tutor: failed attempts to render the problematical figure of the half-man half-horse. He had always been too scared to ask Chiron the story of his birth, as afraid as he was fascinated to broach the subject that the boys all gossiped about in awe. He knew the truth now, of course, about how Chiron's father was **Cronus**, the king of the Titans, and his mother was an Oceanid, Philyra; of how Cronus's wife, **Rhea**, had caught the two in the throes of lust, whereupon Cronus turned himself into a horse, filling the sea nymph's belly with the crossbred Chiron before fleeing the scene and the wrath of his wife.

What howling agony his teacher had suffered, thought Asclepius amid the debris of his school days. Hidden away in his cave, unable to heal that weeping wound. Of course it wasn't Hercules's fault that an arrow of his had accidentally lodged itself in his old tutor's knee during a skirmish with some rowdy centaurs. How could he have known, when, years earlier, he defeated the Hydra and dipped his arrows in her blood, that the poison that clung on to their tips would later find itself coursing through Chiron's veins? How could he have stopped what fate had in store? Nor was it his own fault (though he would suffer agonies of guilt) that Chiron's divine descent meant that although injured, he couldn't die, and so was unable to be released from the constant pain that Asclepius could do nothing to assuage. Mercifully, **Zeus** had allowed Chiron to give his immortality to **Prometheus**, saving the tortured hero from Aquila and letting the kind centaur die in peace.

Asclepius saw something shining at the bottom of the box. It was a medal: an award for "endeavor." He remembered the day he had won it. The clearing in the forest where the games had been held. How he had stood amid the other boys, clasping on to the unwieldy bow that was too big for his body, and looked out to the trees where his arrows had scattered without success. He had been ashamed then, to receive this stupid prize, to have tried so hard and still to have failed; and he was ashamed now.

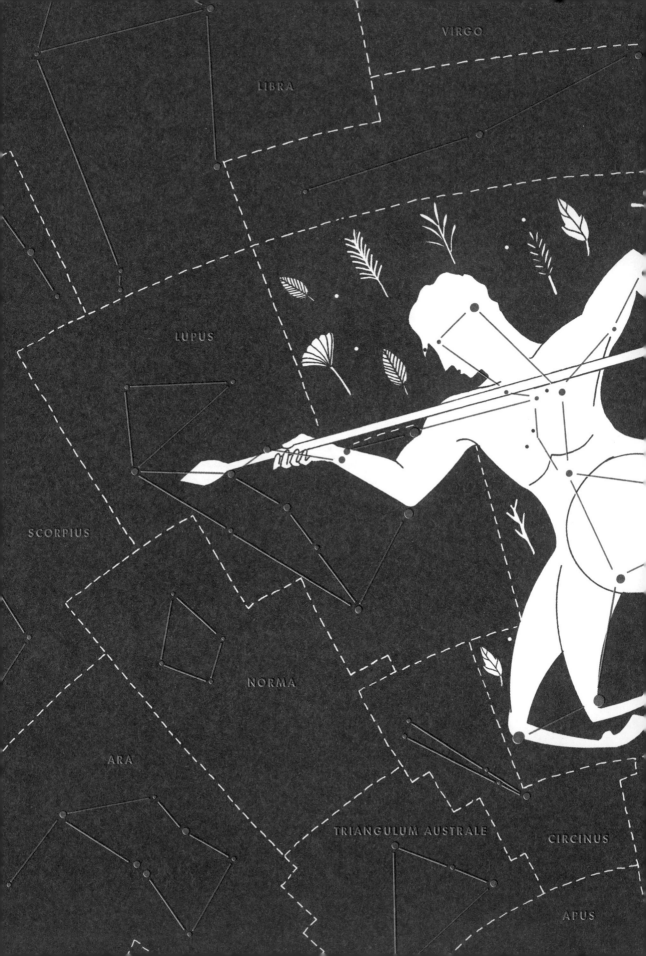

VIRGO

LIBRA

LUPUS

SCORPIUS

NORMA

ARA

TRIANGULUM AUSTRALE

CIRCINUS

APUS

CORVUS

CRATER

HYDRA

ANTLIA

VELA

CRUX

MUSCA

CARINA

0 1 2 3 4 5

MAGNITUDE

CEPHEUS

CEP/CEPHEI, THE ETHIOPIAN KING

RANK IN SIZE: **27**
ASTERISMS: **NONE**

I CANNOT TELL my grief; and yet there you see it, in the cinema of the stars, a shameful family saga played endlessly in the heavens' eternal picture show. I was king of Joppa. Husband to the vainglorious Cassiopeia and father—though I deserve not the name—to the most beautiful of girls, my Andromeda. Nowadays my stars are not so bright and it will take all of your squinting, brain-squeezing imagination to trace my figure in the sky. I'm never the one anyone remembers in the story, in any case. Unlike my wife, whose five most brilliant stars form a famous W—for woman, witch, what have you. Perhaps I was too weak. It's true that my queen had me wrapped around her little finger—but what a beautiful little finger it was . . .

Yet one day, when Cassiopeia was combing her thick, dark locks, her boasts got out of hand. Looking into the gilt mirror before her, she declared that she was more lovely even than **Poseidon**'s fair daughters, the **Nereids**, those extraordinarily beautiful water nymphs. The god of the seas was enraged. He struck his trident on the wine-dark waves and unleashed the full power of his fury. First the lands were flooded by the violence of the water he stirred up, and then there arose from the deep the heinous sea monster Cetus. This monster wreaked havoc on my kingdom. I went to the Oracle of Ammon to ask how to save my people from its jaws, but the answer broke my heart in twain. I must sacrifice my daughter to this hungry fiend or let my subjects face its limitless wrath.

So I took Andromeda to the water's edge and—and—I chained my own daughter to a rock.

Her terrified screams and thrashing limbs still torture me in my dreams. The monster loomed out of the sea and set its foul sights on my darling girl. Her story, thank Zeus, does not end badly. Journeying home from slaughtering the Gorgon **Medusa** was Perseus, who overheard her piercing cries and was as entranced by her beauty as I was by her mother's. Promising to slay the monster in return for her hand in marriage, he found a way to outwit it. He hovered above Cetus with the Sun behind him, and as the beast lurched toward Perseus's shadow, the man himself thrust his sword into its flesh, rescuing Andromeda and carrying her away on his flying steed, **Pegasus**.

She has not, of course, forgiven her mother or me. My punishment is my grief, and the sky's perpetual reminding me of it. Her mother is condemned to circle the Celestial Pole, spending part of the year hanging upside down, legs apart, in a vulgar and shameful posture designed to punish her pride.

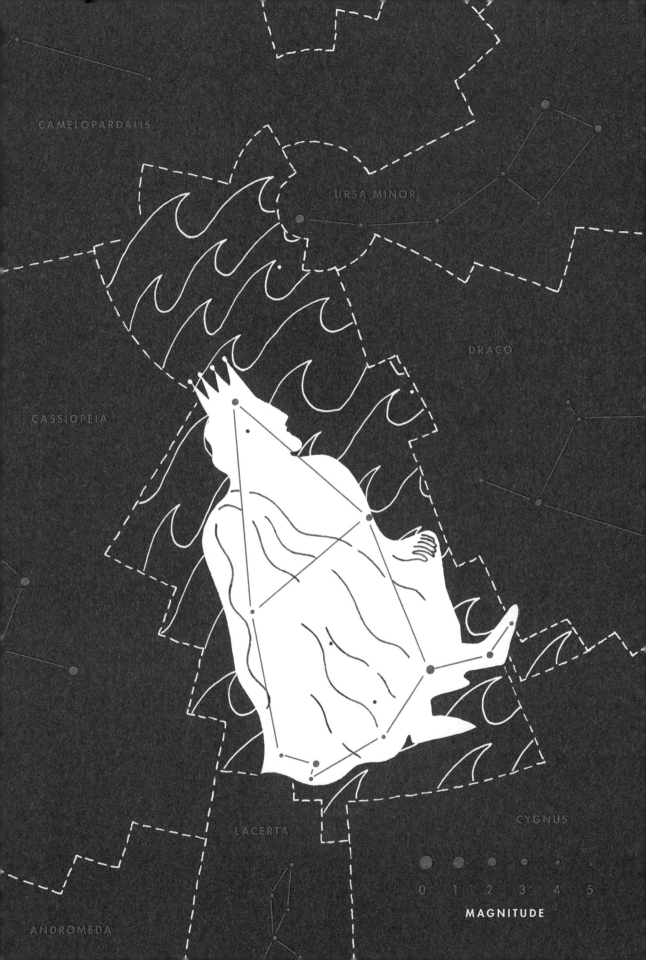

URSA MINOR

DRACO

CASSIOPEIA

CYGNUS

LACERTA

ANDROMEDA

· · · · · · ·
0 1 2 3 4 5

MAGNITUDE

CETUS

CET/CETI, THE SEA MONSTER (WHALE)

....

RANK IN SIZE: **4**
ASTERISMS: **THE HEAD**

....

AN EERIE QUIET, with the muffled noise of seaside shrieks just audible. Long, pretty legs treading water under the sea. A splashing on the water's surface as the sound jumps to full holiday volume. Then the eerie quiet again, and the legs.

You've seen this before.

An innocent girl thrashing about in the shallows, children playing in the waves. Inflatable rafts, excited shrieking, a super-eight summer scene.

You've seen this before.

Deep, deep from the ocean's obscurity, a monster heaves its massive bulk toward the light. The fish dart out of its way. The water shivers cold in its wake. An eerie quiet, and a splashing of legs.

You've seen this before.

What is this stinking creature that slithers up from the celestial depths? That has barnacles on its back and ancient seaweed on its gills? That skulks the oceans in undying greed? That has terrorized the shores, decimated the people, brought out the heroes since time immemorial?

Is it **Tiamat**, the Babylonian beast of primordial chaos? No: she was slain by the sky god **Marduk**. He flew to her on a white horse, and from her slaughtered body were made the heavens and the Earth.

Is it the monster from whose frothing jaws Perseus (flying on winged sandals, and not, as so many storytellers say, the customary hero's steed) saved the chained Andromeda? No: Perseus tricked that foul creature Cetus with his shadow and ran his sword through its scaly flesh.

Is it the dragon in the Libyan lake to which a desperate king was about to feed his beautiful daughter, in a last bid to rid his country of its curse, just at the moment when St. George rode by? The valiant knight saved the princess and slew the vile creature with his Ascalon (the mighty sword whose name, years later, another English patriot, Winston Churchill, would use to christen his own aircraft).

Is it Moby Dick?

Who is this fishy fiend? This faceless monster that has ravaged our shores and our imaginations for millions of years. Who swims up from the dark depths of our minds and attacks us in our dreams. Who we see not just in our nightmares but in the stars.

Legs treading water; children shrieking; the sound of—is it a cello? is it a horn? (no, it's a tuba)—alternating ominously between two notes; something moving slowly upward, aiming for the legs.

I think you know who Cetus really is . . .

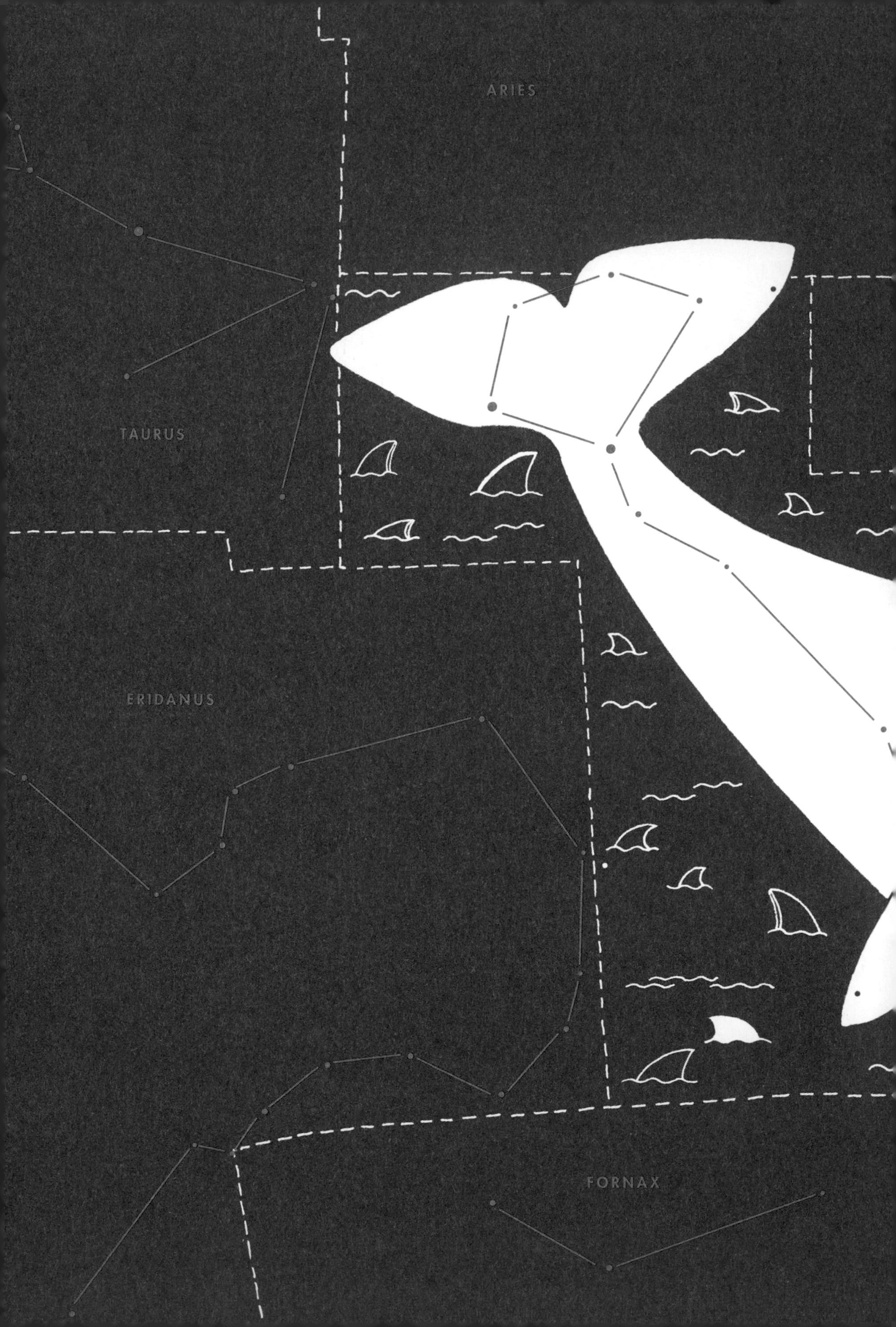
ARIES
TAURUS
ERIDANUS
FORNAX

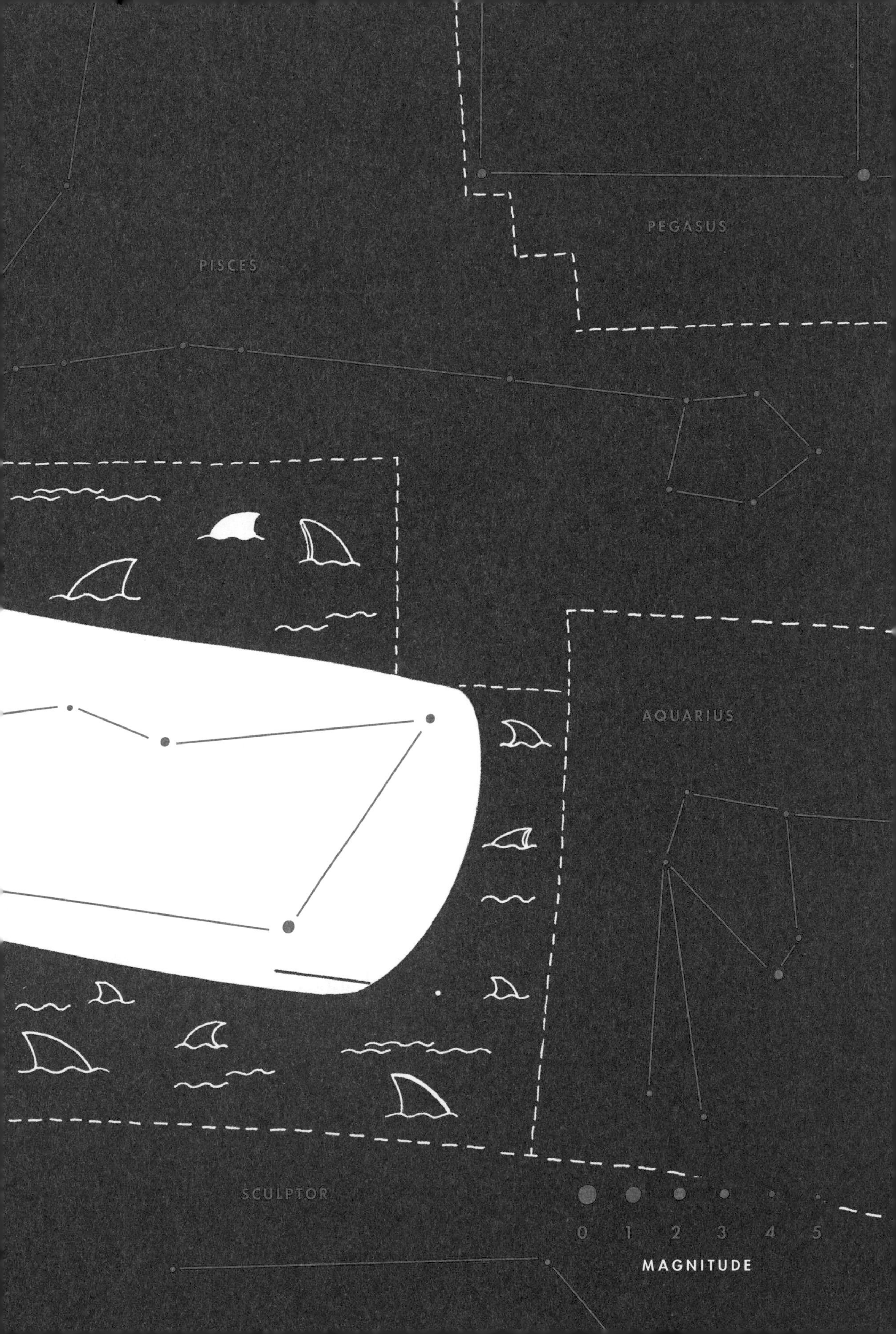
PEGASUS
PISCES
AQUARIUS
SCULPTOR
0 1 2 3 4 5
MAGNITUDE

CHAMAELEON

CHA/CHAMAELEONTIS, THE CHAMELEON

RANK IN SIZE: 79

ASTERISMS: NONE

IT WAS NOT a year too soon when I was finally honored in the stars. I had spent millennia shifting about the night, lighting up as a Chinese tortoise here, a Navajo coyote there. I was particularly good as a Boorong Malleefowl, if I do say so myself.

Indeed, before those kindly Dutch cartographers gave me my own place in the Celestial Sphere, I had been treated really rather badly by your kind. It was that big-nosed Roman meddler of yours, **Ovid**, who started it all—as he did so many mistruths. (On that note, a word of warning: don't get Andromeda going on the subject, whatever you do.) Once Publius Ovidius Naso had waxed lyrical about my eating air and changing color, they were all at it. I can't deny that I enjoyed the attention. As Oscar Wilde (whose capricious nature I of course admire) once said, "There is only one thing in the world worse than being talked about, and that is not being talked about." But did old Will need to be quite so obscure when getting that inky adolescent Hamlet to use me as a metaphor?

King Claudius: *How fares our cousin Hamlet?*

Hamlet: *Excellent, i'faith; of the chameleon's dish: I eat the air, promise-crammed: you cannot feed capons so.*

I've had centuries to ponder the words the young prince spits out at his usurping uncle and still have no idea whether I'm getting a good write-up.

Then there was that tubercular teenager **John Keats**, co-opting me as one of his tribe: "What shocks the virtuous philosopher, delights the camelion Poet." Fair enough, I have scant regard for the opinion of your moral philosophers, but I am not sure I hold your braying bards in any higher esteem. I was most hurt by that young Romantic's insinuation that I have no identity to call my own: "It has no self—it is every thing and nothing—It has no character—it enjoys light and shade; it lives in gusto, be it foul or fair, high or low, rich or poor, mean or elevated." As those worthy explorers **Keyser** and **de Houtman** discovered when they landed in Madagascar in 1595, my ability to camouflage myself in hostile environments is proof of my idiosyncratic skill, my peculiar singularity, not my clonelike inconsequence. Perhaps you humans should look to yourselves, shuffling around your offices in your gray suits, before you cast aspersions on my imaginative flair.

And while I'm at it, I do not appreciate the moniker you Australians use to describe me (I am most certainly not, and never have been, something as drab and domestic as a Frying Pan); nor am I overjoyed by you stargazers in China who nickname me the Little Dipper.

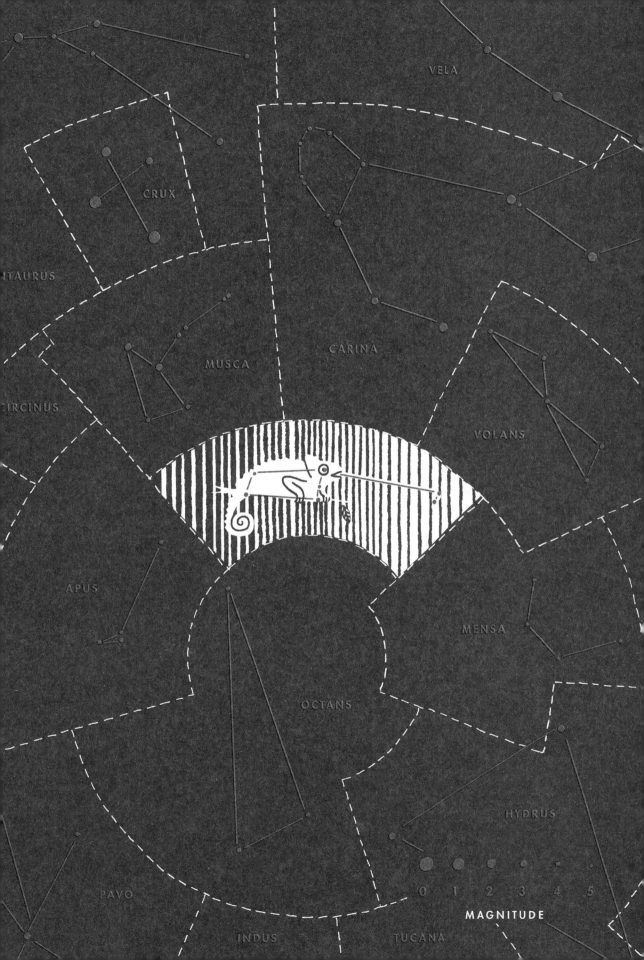

CIRCINUS

CIR/CIRCINI, THE COMPASSES

....

RANK IN SIZE: **85**

ASTERISMS: **NONE**

....

Le silence eternel des ces espaces infinis m'effraie—
The eternal silence of these infinite spaces frightens me.
Blaise Pascal (1623–62)

YET ANOTHER of the figures created by the French astronomer **Nicolas Louis de Lacaille** to commemorate the tools of the mathematical and scientific arts, Circinus represents the pair of compasses used by surveyors and sits beside another of their tools—Norma, the surveyor's level—in the night sky. This apparently insignificant constellation—which was shown in 1970 to contain an entire galaxy previously hidden to human eyes—does not seem so to me. Ironically, it has the absolutely inverse effect: its relative tininess serves only to terrify me with the vast and panic-inducing unknowability of space. Just trying to conceive of the size of the perfect circle that this celestial instrument eternally draws onto the universe awakens parts of my mind that I prefer to leave snoozing in a dull stupor.

The poet John Donne (1572–1631) was someone else troubled by the metaphysical implications of scientific discovery. Much of his writing grapples with the new and mind-exploding developments in astronomy that were happening all around him during the Renaissance. A tortured soul in the first place, this Roman Catholic dandy turned Anglican dean of St. Paul's Cathedral seems very unsure of what to make of the findings of **Nicolaus Copernicus**, **Tycho Brahe**, **Galileo**, and **Johannes Kepler**. Historians have argued as to whether his position was one of religious skepticism or intellectual admiration, but his feelings were clearly as complicated and ambiguous as much of his poetry.

One thing he was certain about was how much he loved his wife. In perhaps my favorite of his—or indeed any—poems, "A Valediction: Forbidding Mourning," a pair of compasses is used as a metaphor for the one thing that can quell my fear of the eternal silence of infinite space: love.

Our two souls therefore, which are one,
Though I must go, endure not yet
A breach, but an expansion,
Like gold to airy thinness beat.

If they be two, they are two so
As stiff twin compasses are two;
Thy soul, the fixed foot, makes no show
To move, but doth if th' other do.

And though it in the centre sit,
Yet when the other far doth roam,
It leans and hearkens after it,
And grows erect, as that comes home.

Such wilt thou be to me, who must,
Like th' other foot, obliquely run;
Thy firmness makes my circle just,
And makes me end where I begun.

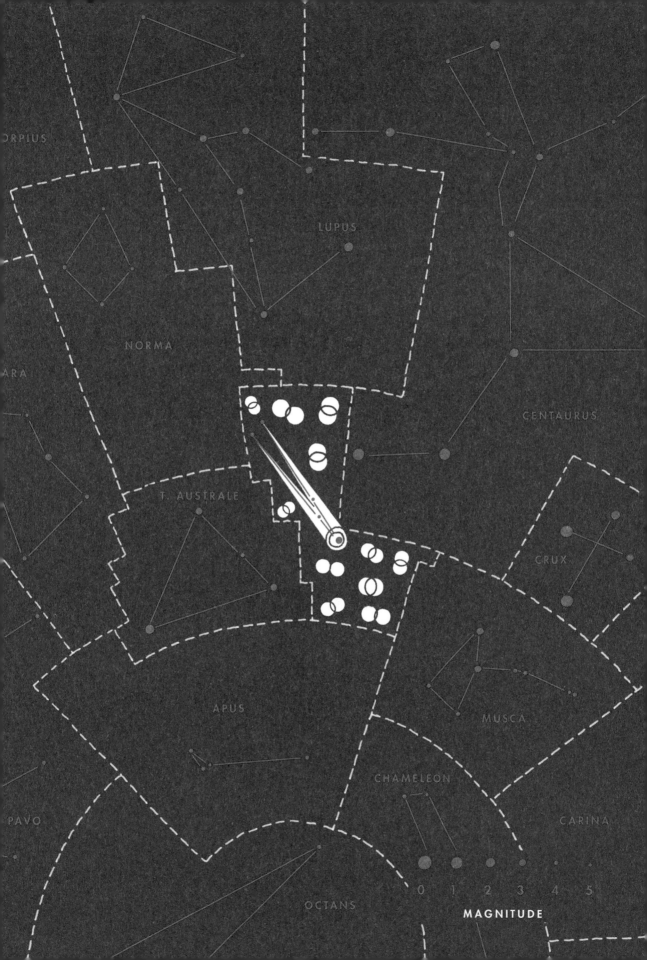

COLUMBA
COL/COLUMBAE, THE DOVE

RANK IN SIZE: **54**
ASTERISMS: **NONE**

BEFORE THE Spanish Civil War, the painter Pablo Picasso was, according to his dealer Daniel-Henry Kahnweiler, "the most apolitical man" he had ever known. But the horror of that conflict resulted not only in one of his most famous paintings but a lifelong commitment to the communist battle against fascism, and the struggle for liberty and peace. The chaos of suffering depicted in his eponymous *Guernica*, painted in immediate response to the 1937 Nazi bombing of that Basque town, shows an imagination uprooted by the horror of war. Full of symbol and metaphor, its figures are both mythical and material—a **Minotaur** and a lightbulb equally potent in the devastation they allude to. Picasso, who was fascinated by ancient and classical legend, was clearly an artist in whose vision the stars had been realigned.

When he was a boy, his family home was full of doves, and his father taught him how to paint them. Later in life there were always doves cooing around his own studios and home. In 1949, the poet and editor Louis Aragon visited his studio and chose a beautiful lithograph of a white dove as the image for the poster of the inaugural communist World Peace Congress in Paris. Picasso later simplified this original—a soft and delicate likeness of a pigeon given to him by his friend and great rival Henri Matisse—into a simple line drawing, which has now become one of the world's most famous and recognizable emblems of peace. The night before the conference, Françoise Gilot gave birth to a baby girl, Picasso's fourth child. They called her Paloma—the Spanish word for dove.

The dove in the stars flies behind **Jason** and the Argonauts' vessel *Argo*, the ship it once guided through the sliding doors of the Symplegades—the clashing rocks, which crushed any vessels that passed between them. But it was first drawn as a constellation on a celestial globe created by the Dutch astronomer and cartographer **Petrus Plancius**, very much with the biblical, olive-branch-bearing dove in mind. He even renamed the nautical Argo Navis constellation as Noah's Ark. But it is the Hellenic story that has stuck and, like the bird in this tale, both the named stars in Columba are harbingers of good news. Its brightest is the blue-white, third-magnitude Phakt, named after the Arabic for "ring dove"; while β Columbae is a yellow star called Wasn, meaning "weight." Is it so terribly naive to hope that one day, when more than enough people will have looked up in suffering at this small, celestial dove, we will at last have peace?

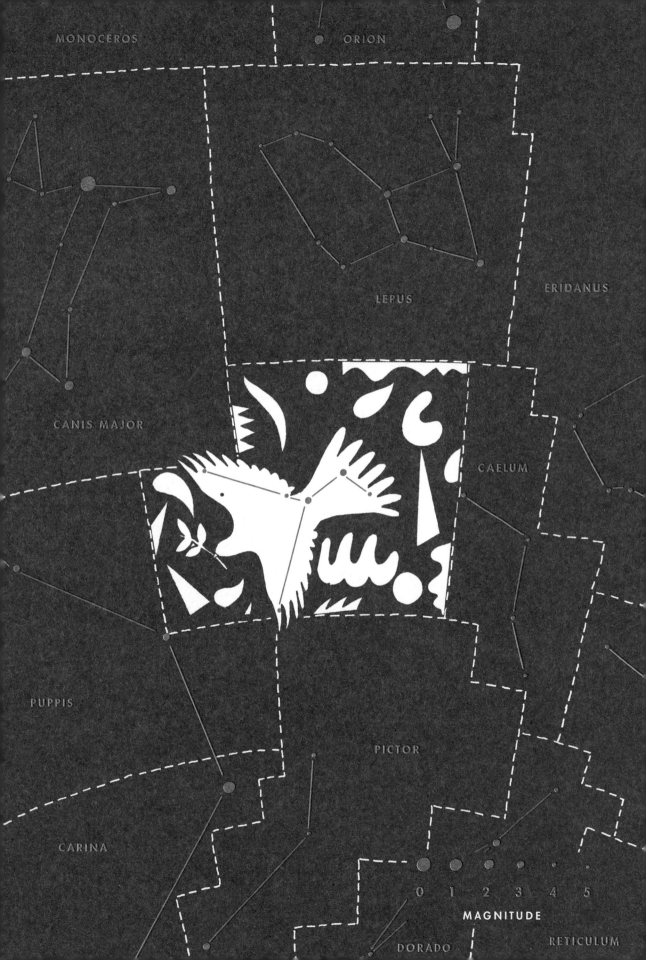

MONOCEROS

ORION

LEPUS

ERIDANUS

CANIS MAJOR

CAELUM

PUPPIS

PICTOR

CARINA

RETICULUM

MAGNITUDE

0 1 2 3 4 5

DORADO

COMA BERENICES
COM/COMAE BERENICES, BERENICE'S HAIR

RANK IN SIZE: **42**
ASTERISMS: **NONE**

This Nymph, to the Destruction of Mankind,
Nourish'd two Locks, which graceful hung behind
In equal Curls, and well conspir'd to deck
With shining Ringlets the smooth Iv'ry Neck.
Love in these Labyrinths his Slaves detains,
And mighty Hearts are held in slender Chains.
Alexander Pope, *The Rape of the Lock*

"THAT PUFFED-UP, powdered ponce. All airs and graces. Somewhere along the line he seems to have forgotten that he's a poet and not the head of the Holy Roman Church . . ."

The barber paused mid–razor stroke and tutted, as if in explanation. The young gentleman being shaved flinched in apprehension as the portly locutor resumed his precarious task.

"That's how the whole thing started, apparently. Some sort of love tryst between a bunch of posh Catholics—Lord something-or-other and Lady this-or-that—him stealing a lock of her hair, the la-dee-dah families getting their knickerbockers in a twist, and the old rhymester himself being brought in to write some nonsensical verses to patch the whole quarrel up. At least, that's the story one of my *society* clients told me—just between you and me!"

He winked grotesquely into the mirror.

"It's all right for fancy-pants Alexander—he wears a wig. You try refashioning a head of hair once it's been hacked into by a lovesick Arabella, Camilla, or Belinda of the beau monde."

He gave another theatrical tut.

"It started innocent enough—one airy-headed baroness came in asking me to cut off her finest lock—'like in the *Heroik Poem*,' she said, whatever that is when it's at home. Er—hang on, sir, just a little nick . . ." he said hastily, dabbing at the young man's increasingly pallid cheek.

"But then a marchioness came along and had half her curls lobbed off and now the whole of London's gone topsy-turvy for it. Apparently, the bombastic old bard based the sodding limerick on an ancient legend about some old biddy who cut off her hair and got it put in the stars."

"Queen Berenice of Egypt," the young man interrupted, "a genuine historical figure. She placed her exquisite locks as a votive offering at the temple of **Aphrodite** in thanks for her husband's safe return from battle. When, to the horror of the court, the regal tresses disappeared the next day, Conon of Samos, a mathematician and astronomer in Alexandria, appeased the king's rage by pointing aloft to a brilliant cluster of stars. Berenice's hair, he explained, had ascended to the celestial domain."

"Well, something or other along those lines, yes," the barber replied. "It's all Greek to me."

CANES VENATICI

URSA MAJOR

LEO

BOÖTES

VIRGO

MAGNITUDE

0 1 2 3 4 5

CORONA AUSTRALIS

CRA/CORONAE AUSTRALIS, THE SOUTHERN CROWN

RANK IN SIZE: **80**
ASTERISMS: **NONE**

KATIE WAS NOT beautiful, so she decided she would be clever. We can presume then that this studious girl, born to an Anglican family at Encounter Bay in Southern Australia in 1856, would have made sure to know her Greek myths as well as her Bible. She may have gleaned some of them from her reading of the Roman poet **Ovid**. Of course her mother, Sophia Field, may not have taught Latin in the lessons that she gave Katie and the Aboriginal girl who had saved her daughter from drowning. But whether Mrs. Field introduced them to the pleasures of the classics, we can be sure that the ever-curious Katie would have tugged on the shirtsleeves of her pastoralist father, Henry Field, and asked him to bring his dusty *Metamorphoses* down from the bookshelf.

She would have sat on the stairs and read of **Jupiter** and Semele, who had grown big with his seed. (Out of instinct, she never asked her father exactly what this meant—in any case, she considered, he raised livestock, not crops.) She would have read of Jupiter's wife, **Juno**, upset and betrayed, who dressed up as Semele's old nurse and asked her how she could be sure it was really Jupiter who had made her grow so big. Katie would have blushed and run upstairs to her bedroom to finish the final part of the story.

The counterfeit nurse persuades Semele to ask the god to make love to her just as he does to Juno. The young mortal, unable to withstand the heat of Jupiter's lust, erupts into flames in his embrace. Her unborn baby is ripped out of her womb and sewn into Jupiter's thigh to fully gestate. Years later, the boy—who has grown up to become **Bacchus**, the god of wine—goes to the underworld with a gift of myrtle to retrieve his mother's soul. The gods agree that Semele can join them on Mount Olympus, and Bacchus places a wreath in the stars to honor his mother. Ever after, the god's followers (famous, Katie must have read with saucer-wide eyes, for their debauched, ecstatic rites) wore crowns of myrtle.

Years later, Katie—who had grown up to become K. Langloh Parker, a wife and lay anthropologist—had all but forgotten this story. For some time, she had been discovering an altogether different collection of myths. Sitting around the campfire, she had gathered the legends of her Aboriginal neighbors in the remote part of the outback where she and her husband now lived. To her, the stars of Corona Australis were no longer Semele's wreath but a boomerang; and although Mrs. Parker may not have known this, the only example of a "connect-the-dots" asterism in Aboriginal astronomy.

SAGITTARIUS

SCORPIUS

TELESCOPIUM

ARA

PAVO

NORMA

● ● ● ● ● ●
0   1   2   3   4   5

MAGNITUDE

CORONA BOREALIS
CRB/CORONAE BOREALIS, THE NORTHERN CROWN

RANK IN SIZE: **73**
ASTERISMS: **NONE**

HOW TO MAKE A STAR CROWN

STRING TOGETHER a glittering circlet of seven stories and place it the northern sky, between Boötes and Hercules. Fix your wreath of tales—in this order—to these seven stars: Iota (ι), Epsilon (ε), Delta (δ), Gamma (γ), Alpha (α), Beta (β), and Theta (θ) Coronae Borealis. You'll need to use superglue (Elmer's won't be strong enough), so be sure to do this with adult supervision. Remember to save your best tale for Alpha Coronae Borealis: this blue-white star known as Alphecca (meaning "bright one of the dish") in Arabic, or Gemma (the Latin word for jewel) will be your crown's most sparkling gem. You can pick any stories you like to adorn your celestial diadem, but here are seven that have been stuck to its stars before.

1 The Chukchi Legend of the Polar Bear's Paw.

2 Caer Arianrhod: the Fortress of Arianrhod, the Lady of the Silver Wheel. A Welsh legend in which the daughter of the goddess Dôn steps over a magical staff and instantly "drops" a yellow-haired boy, Dylan—a sea spirit who flies away to sea. And also . . . a mysterious, amorphous lump of flesh, which her brother Gwydion hides in a chest at the foot of his bed. It morphs into a boy who grows twice as fast as a normal child, and when he is four years old his uncle takes him to Arianrhod's fortress by the sea. In shame and disgust his mother places a series of curses on him, that he and Gwydion must plot to overcome.

3 The Middle Eastern Beggar's Dish.

4 Ariadne's Crown: a Grecian tale of woe in which the beautiful daughter of **King Minos** of Crete falls in love with the handsome Theseus, who has come from Athens to kill the **Minotaur**. She gives him a ball of golden twine so that he can escape the beast's labyrinth; they marry, but then he abandons her on the island of Naxos. Dionysus spots the grieving beauty, weds her in an instant, and gives her a jewel-encrusted crown fashioned by **Hephaestus**, the god of fire, himself.

5 Kilu's Boot: a Siberian story from Koryak.

6 A Tale of Chinese Entrapment, featuring T'ien-lao, the Celestial Prison; Lien-ying, the Endless Enclosure; T'ien-wei, the Celestial Jail; and last but not least Guansuo, the prison for working-class miscreants.

7 The Bornean Fish.

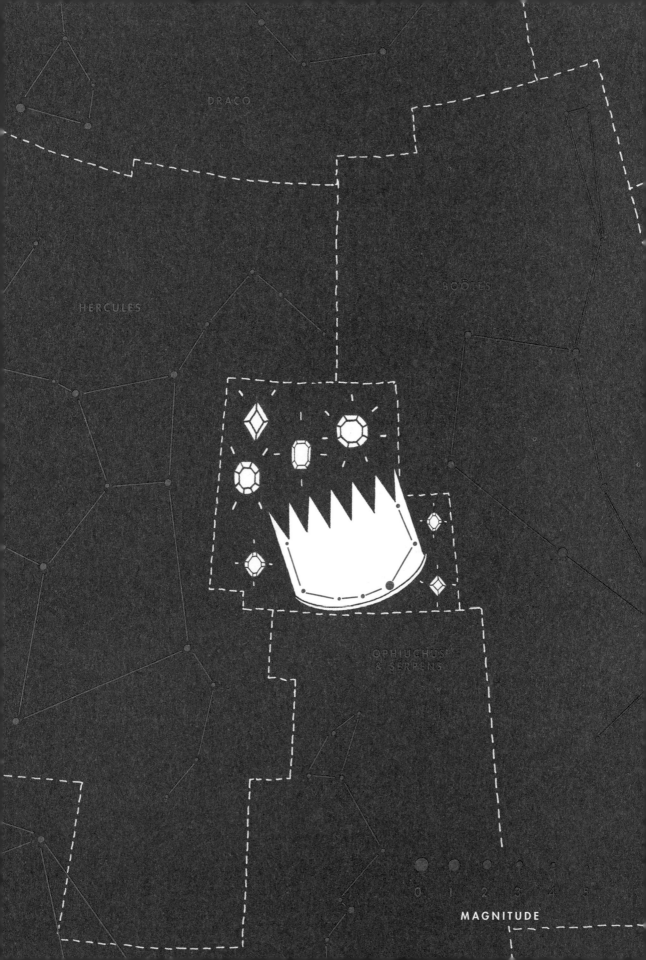

DRACO

HERCULES

BOÖTES

OPHIUCHUS
& SERPENS

MAGNITUDE
0 1 2 3 4 5

CORVUS

CRV/CORVI, THE CROW

RANK IN SIZE: **70**
ASTERISMS: **THE SAIL**

WE ARE NOT so different, you and I. Intelligent, curious, destructive. We manufacture tools, play sports, and know our enemies by their faces. We understand that there is more than just the present, and so we communicate and plot. Our societies are structured and sprawling, littering waste and disease. We are omnivorous scavengers that all the other animals hate.

But is my collective noun for you a murder of men? Do my myths blacken your name? Are you as cunning, malevolent, and—worst of all—stupid in our myths as we are in yours? My squawk is so ugly to you that you have invented fanciful legends to explain how I lost the power of song. You tell a story that makes us laugh. That one of your gods (and why on earth do you have so many and why can't you agree about them?) **Apollo**, who you see as my master (if only you understood the irony of that), once sent me off with his chalice to fetch the waters of life. In it I am, of course, stupid and punished and all this is commemorated by you in your human and erroneous apportioning of the sky: a story you can read in the constellations you call Corvus and Crater that sit on Hydra's back.

Let me tell you the real story about the stars.

Once upon a crow, there was a good time; when the land was full of flesh to feed on, and the soil was free of poison. The sun would last all day and there was no such thing as night. But one day Great Mother Crow and Great Father Crow had a terrible row. Great Father Crow had eaten the last worm on her plate of skeleton and then refused to stroke her feathers and clean their nest. To punish her husband for abandoning his husbandly duties, Great Mother Crow placed all the world in darkness so that no one could see one another, and the whole crow world was plunged into chaos. Parents stepped on their children's heads, and lovers collided into trees to their deaths. Besieged with grief and regret for what he had done, Great Father Crow tried to make amends. He pecked and pecked at the darkness to free all the light he could from the sky, but as hard as he worked he could pierce only small holes in the all-enveloping black. Finally Great Mother Crow relented. She gave the crows back the Sun. But she kept it from them for half of the day to remind husbands always to obey their wives. And this is why you humans have your night.

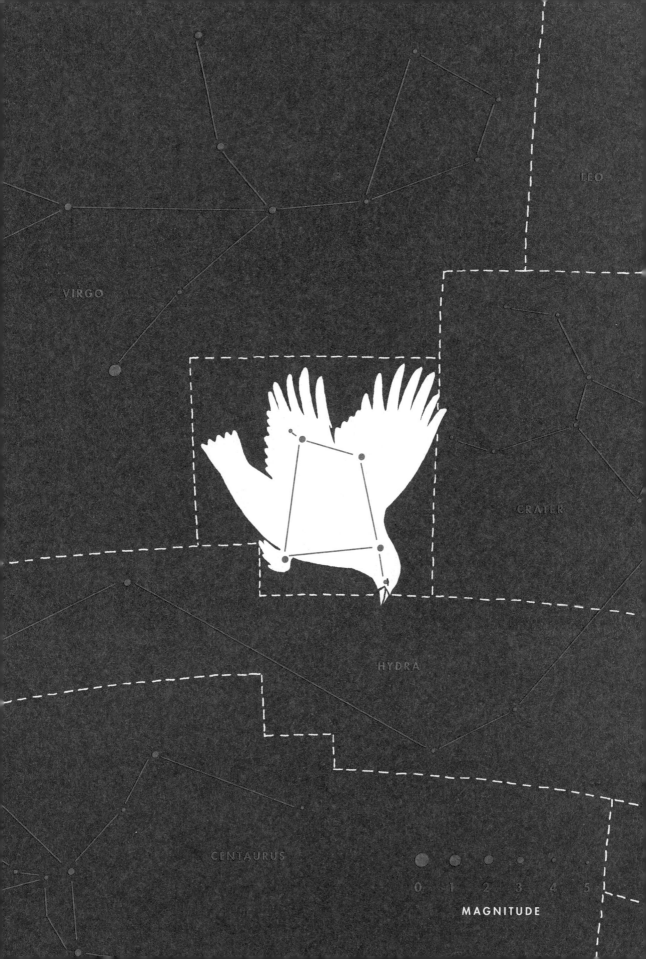

LEO

VIRGO

CRATER

HYDRA

CENTAURUS

0   1   2   3   4   5

MAGNITUDE

CRATER

CRT/CRATERIS, THE CUP

RANK IN SIZE: **53**
ASTERISMS: **NONE**

WINE AND WATER. Water and wine. This sacred drinking vessel was one of **Ptolemy**'s original forty-eight constellations and has since metamorphosed into a Spanish amphora, a Persian wine cup, and a German bucket. But it is best known as a *krater*—an ornate, double-handed chalice the Greeks used to mix water with wine. It is part of a celestial morality tale that straddles the largest constellation in the sky, the many-headed monster, Hydra. Sitting on the back of this galactically-size female water-snake are Corvus, the Crow or Raven, and Crater, the cup that **Apollo** gave to his sacred bird to fetch the waters of life.

Now, the quick-witted among you may remember that Crow wasn't in his master's good books after he had delivered to the god news of his lover **Coronis**'s infidelity. In fact, so full of rage was the envious Apollo that the mother of his son Ophiuchus had betrayed him—and with a *mortal*, too—that he had cursed the snow-white bird and singed all its feathers black. So you would have thought that our winged hero, usually so clever and full of tricks, would have been on his best behavior when he was tasked with filling up this holy beaker.

He set off earnestly enough, holding Apollo's chalice in his claws and scanning the land beneath him for the precious spring. Crow beat his wings strongly: well he knew that his master wished to make a sacrifice to mighty **Zeus** and that the sacred liquid must be brought back swiftly to satisfy the loud-thundering god. He had not been flying long when a fig tree appeared in his view. Surely he could be allowed just a short diversion? Settling on its branches, he inspected the tree's plump fruits. Tantalizingly, they were not quite ripe. But what a shame to leave such fine figs untouched! I will wait, he thought, till their green skins turn purple, and then I will fly straight to the spring and fetch the holy waters.

By the time Crow was sated and his beak gloriously stained, he had completely forgotten his errand. It was only as he lifted himself off the branches with a groan—which was hard to do with a belly so heavy with indulgence—that he remembered his task with a shock. He raced to the spring to fill up his cup and sped back home to his master. Apollo did not fall for the excuse Crow gave that he had been delayed by a water-snake blocking the source. He cursed the lazy servant and set him in the stars on Hydra's back, just out of the reach of Crater, in eternal and tantalizing thirst.

VIRGO
LEO
SEXTANS
CORVUS
HYDRA
ANTLIA
CENTAURUS
0 1 2 3 4 5
MAGNITUDE

CRUX

CRU/CRUCIS, THE (SOUTHERN) CROSS

RANK IN SIZE: **88**
ASTERISMS: **NONE**

AN ACROSTIC CROSSWORD

ACROSS

1 T he Maori of New Zealand call this constellation *Te Punga*—the anchor of the Milky Way

2 H ailed by patriotic Australians who see themselves as its sons and daughters

3 E arth's precession on its axis means that this Christian symbol hasn't appeared
in the northern hemisphere since the era of Christ, when it was just visible in Jerusalem

4 S tars in which Crosby, Stills & Nash found consolation for the brokenhearted in their
1982 song

5 O to be beneath this, sang Patti Smith, under which even the gods get lost

6 U nder this proudly stand the singers of the Australian cricket team's victory song

7 T he Tongans call it *Toloa*—a duck with a wounded wing flying south after two
men threw a stone at it

8 H eralded as a symbol of the southern nations since colonial times, it features
on the flags of Brazil, Samoa, New Zealand, and Papua New Guinea

9 E leventh-century Arabic astrologer Abu al-Biruni recorded a southern asterism
visible in parts of India and known as *Sula* ("the Crucifixion Beam")

10 R egarded as two giraffes—*Dithutlwa*—by the Tswana people of Botswana:
the stars Acrux and Mimosa form a male, and Gacrux and Delta Crucis a female

11 N avigational aid for sailors in the southern hemisphere: pointer to the
South Celestial Pole

DOWN

1 C ontains the Coalsack Nebula, a cloud of dark gas and dust 400 light-years away,
seen by Australian Aborigines as the head of a massive evil emu

2 R epresented by **Ptolemy** and the ancient Greeks as being part of the
constellation Centaurus

3 O n a Medici-funded voyage to the Indian Ocean in 1515, Andrea Corsali was
the first European to fully map "this crosse . . . so fayre and bewtiful"

4 S een by Beatrice and Dante in the *Divine Comedy*; according to the poet its four stars
represent the cardinal moral virtues Prudence, Justice, Temperance, and Fortitude

5 S mallest of all the eighty-eight constellations in the sky

CENTAURUS

HYDRA

VELA

CIRCINUS

MUSCA

CARINA

TRALE

CHAMAELEON

VOLANS

APUS

MAGNITUDE

0 1 2 3 4 5

OCTANS

CYGNUS
CYG/CYGNI, THE SWAN

RANK IN SIZE: **63**
ASTERISMS: **THE NORTHERN CROSS, THE SUMMER TRIANGLE**

IT IS STILL on the water, and a cold winter light sharpens the estuary, the fields, and the naked trees beyond, widening the window's depth of field. For the first time I see it: a lamentation of swans.

I wonder which ancient swan it was that beat its wings up from the water to fix itself forever in the sky: which swan is lit in eternal flight through the Milky Way, its wings, neck, and tail crucified in the Northern Cross.

I think of poor **Leda** and shudder. Cold **Zeus** as a swan coming down on her in all his angular horror: the flapping, the beak, the pain. I used to think Yeats said it best:

> How can those terrified vague fingers push
> The feathered glory from her loosening thighs?
> And how can body, laid in that white rush,
> But feel the strange heart beating where it lies?
>
> A shudder in the loins engenders there
> The broken wall, the burning roof and tower
> And Agamemnon dead.

But now I'm not so sure. All these men telling beautiful rapes.

Perhaps the swan is **Cycnus**, the son of **Neptune**, whose inhuman strength resisted the spears of warrior Achilles to the very last. Ted Hughes tells the Trojan battle well: how "the impenetrable neck sinews / Of this supernatural hero" held strong until finally "the full berserk might of Achilles" slew him to the ground. But when triumphant Achilles undid the "corpse's gorgeous armour" he found it empty: Neptune had transformed his son into a swan.

I hope the swan is in fact a different Cycnus, the sad friend of **Phaethon**, whose plaintive mourning is honored in the stars. When brash Phaethon fell from the sun god's burning chariot and plummeted to his death in the muddy depths of the river Eridanus, his oldest and best friend, Cycnus, looked on in helpless horror. He mourned and mourned on the river's bank until his hair turned white and his toes very cold. At last the gods took pity. The hairs on his head sprouted soft and wide and his nose struck forward from his face. His neck grew long, and his eyes shrunk black and he found himself gliding on the water's tide: a swan. He dived to the river's murky bed and brought up his dear friend's corpse to bury in sacred ground.

The lamentation has reached the corner of my view now. A branch floats, lost, in the river as the swans glide coldly downstream.

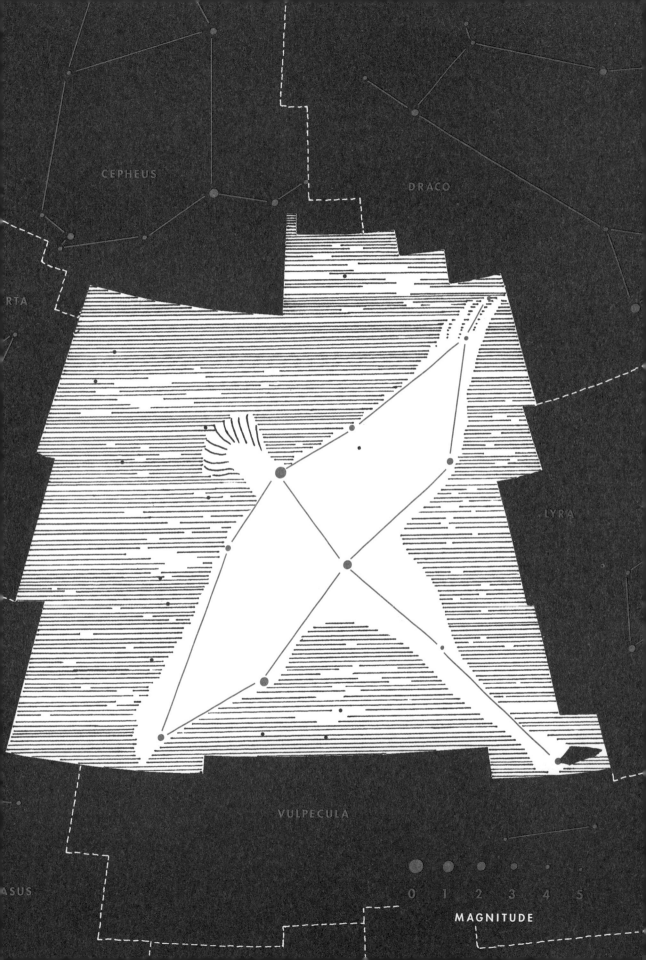

CEPHEUS

DRACO

RTA

LYRA

VULPECULA

ASUS

MAGNITUDE

0 1 2 3 4 5

DELPHINUS
DEL/DELPHINI, THE DOLPHIN

RANK IN SIZE: **69**
ASTERISMS: **JOB'S COFFIN**

THE FIRST TIME I saw a dolphin was on breakfast television: Flipper the friendly dolphin was charming children everywhere with his cheeky grin and delightful chirruping—which, I was recently disappointed to discover, was in fact the doctored sound of a kookaburra. Flipper was ludicrously more intelligent even than we know dolphins to be, chatting away with the show's sandy-haired heroes and saving lives, but this anthropomorphized companion followed an ancient tradition. Dolphins have always been beloved creatures, a sort of marine equivalent to a faithful dog, man's best friend in the ocean.

Greek sailors saw this sacred "fish," leaping eagerly in a ship's wake just as a dog chases a ball, as a servant of **Poseidon**. After the battle of the Titans, when **Zeus**, Poseidon, and **Hades** had overthrown their father, **Cronos**, and divided the world between them, Poseidon became god of the sea and built a glorious palace under the waves. But for all his palace's submarine splendor, the royal bed (carved from the finest coral and plumped with the softest seaweed-stuffed pillows) felt rather empty. Swimming between the reefs in search of a bride he came across the Nereid Amphitrite. Her tresses danced in the ocean currents, and her eyes caught the Sun as it bounced back off the seabed. Poseidon was smitten. Day in, day out, he sent his messengers to her bearing love tokens. He sent a coral handkerchief, a bouquet of phosphorescent flowers, a smooth white pearl from the rarest oyster in the sea. But every gift he sent was met with giggles from Amphitrite and her **Nereids**, so day in, day out, his messengers returned to him, defeated. It was only when Poseidon sent his dolphin to Amphitrite that she finally turned her head. To thank his loyal servant for wooing his bride, Poseidon placed it in the stars, so that now when it plays and dives and splashes its tail, it does so in the heavenly waters.

Amphitrite, however, can still be found under earthly waters. If you dive fifty-five feet under the sea at Sunset Reef, just off Grand Cayman Island, you will see a nine-foot-tall, six-hundred-pound heavy bronze statue of this beautiful mermaid staring up to the light as it glitters at the water's surface.

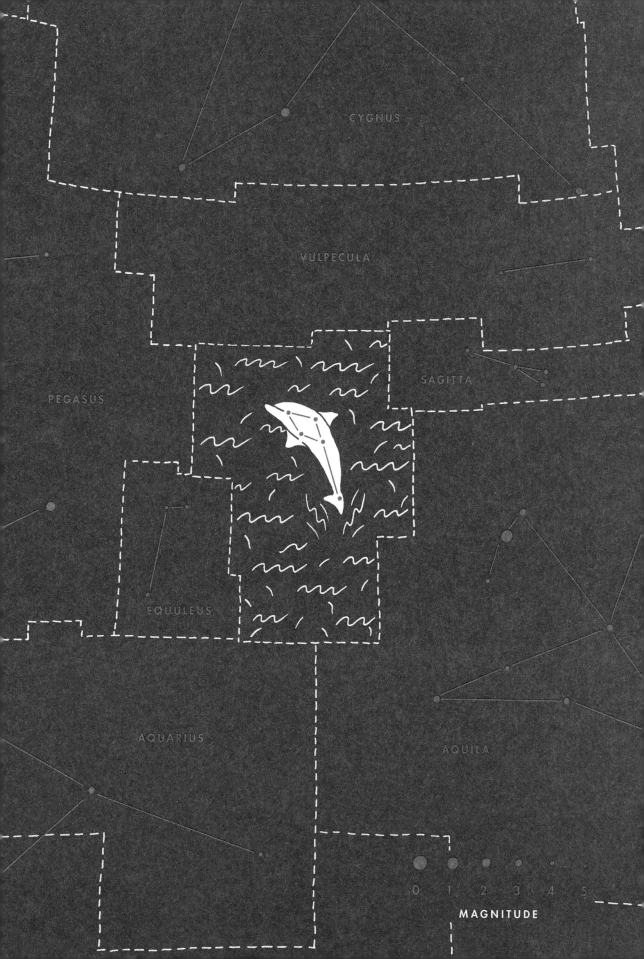

CYGNUS

VULPECULA

PEGASUS

SAGITTA

EQUULEUS

AQUARIUS

AQUILA

MAGNITUDE

0 1 2 3 4 5

DORADO

DOR/DORADUS, THE GOLDFISH

RANK IN SIZE: *72*

ASTERISMS: **NONE**

THE GOLDFISH in the sky is not a memory-poor, bowl-crazed domestic pet of the sort you win at a county fair, but rather a Hawaiian mahi-mahi. Ray-finned, the Dorado (as the Portuguese named it) flashes gold, blue, and green on the surface of the sea. When caught, and exposed to the air on a trawler's stinking deck, it changes color, its bright hues fading with its life until finally it dies, a muted yellow-gray. Like the European navigators who first spotted this exotic fish in the tropical and subtropical waters of the Indian, Atlantic, and Pacific Oceans, it is a highly migratory species and can be found exploring the coasts of the Caribbean, Costa Rica, Africa, Mexico, China, and many others in between.

Placed in the southern stars by those same voyagers, Dorado is relatively faint and undramatic to the naked eye. However, just as the toucan—another of **Keyser** and **de Houtman**'s celestial menagerie—reveals a wealth of treasures when seen through a telescope, so too this unassuming constellation is home to some extraordinary deep-space objects. Most famous is the Large Magellanic Cloud (its partner, the Small Magellanic Cloud is in Tucana). Twenty times as wide as the Moon, and with a diameter of 20,000 light-years, the LMC is the largest satellite galaxy of the Milky Way. You can easily see the faint cloud floating in the night sky without the aid of any astronomical instruments, but when magnified—even just through a pair of binoculars—it transforms into a broad and astonishing bar of stars. A powerful telescope further reveals a cornucopia of nebulae and star clusters in this gas- and dust-rich galaxy.

Elsewhere in the depths of this very unordinary goldfish is the thrillingly named Tarantula Nebula. This star-forming region has long, spindly tendrils of gas like a spider's legs, but at 1,000 light-years across, it would not fit quite so neatly behind your baseboard. At the heart of this stellar arachnid is a dense and massive cluster of 1- to 2-million-year-old blue stars pumping out ultraviolet rays, much as its earthly counterpart pumps out venom to poison its prey. And right at the center of this, among a tight knot of stars, is R136a1: 10 million times brighter than the Sun, and 265 times as heavy, this is the biggest star known to humankind.

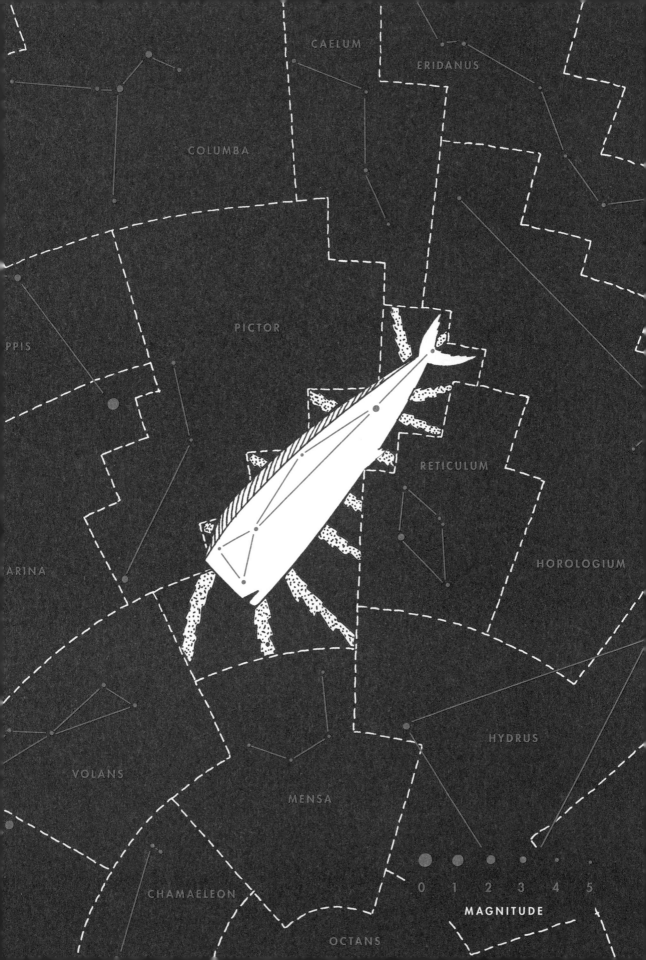

CAELUM

ERIDANUS

COLUMBA

PICTOR

PPIS

RETICULUM

ARINA

HOROLOGIUM

HYDRUS

VOLANS

MENSA

CHAMAELEON

0 1 2 3 4 5

MAGNITUDE

OCTANS

DRACO
DRA/DRACONIS, THE DRAGON

RANK IN SIZE: **8**
ASTERISMS: **THE HEAD, THE LOZENGE**

A SLOW BLACK ire fills my heart. It's there from the first moment I wake in the morning, my scales snagging against the rock of this dark lair, the tip of my tail twitching toward the light. It sits like tar in my innards, as if I have breathed in true sadness and let its vapors ferment.

I don't know how to live in the light among people who are light-stepped and free-souled, and that bitterness has curdled into rage. A rage that I try endlessly to expel in ferocious flames. Mine is the fire of despair; but, after all these thousands and millennia of years, it is indistinguishable from wrath.

I was alive before the sea split from the sky. In the unlit chaos of the big nothing. I remember that void in my bones and can breathe its terror into any human heart. Its emptiness is imprinted in the dry cracks of my coiling skin and its timeless desolation fills the cave I hide in.

With the coming of the light, the good-seekers, the evil-hunters, the kind-hearted, of course they sought me out. The gods who rose up from the depths of that chaos challenged me to battle. But even the gods turned in fear at my might. The humans named me Tiamat, the Serpent, Azhdeha, the Evil One, as if naming things could make them disappear.

I have sometimes been defeated, temporarily of course. Early on, Marduk of Babylon came at me with a bag of tricks—all the goodness the heavens could muster—and propelled me into the stars. Hercules called me **Ladon** and slew me with an arrow. **Athena** caught me by the tail and spun me around and hurled me up into these starry coils. I was frozen and knotted around the Pole Star for years. An astronomer, Thales, clipped my wings and gave them to Ursa Minor.

But mine is an anger that will never cease. A sorrow that will always lurk in the seas of chaos and the skies of your mind. For the darkness of the big nothing is at the heart of things, and my sorrow will spread its ire. Today you have your black dogs. The ancients knew that I am a dragon.

CANES VENATICI

CAMELOPARDALIS

URSA MINOR

BOÖTES

HEUS

HERCULES

CYGNUS

● ● ● ● ● ● ●
0 1 2 3 4 5

MAGNITUDE

LYRA

EQUULEUS

EQU/EQUULEI, THE LITTLE HORSE

RANK IN SIZE: **87**

ASTERISMS: **NONE**

TRAILING CLOSE to the winged steed Pegasus are a dolphin and a little horse: Equuleus and Delphinus. But all we can see of Equuleus is its head. The little horse is the second-smallest constellation in the sky and many have supposed that the faintly starred foal is Celeris, the swift horse that Pegasus sired and that **Hermes** gave to one of the Gemini twins. Some say that Hermes gave Celeris to **Castor**, the twin brother famous for taming horses, and others that he gave it to **Pollux**, the bare-knuckle boxer of the two. But in any case, they're all wrong. For the foal that rises just before Pegasus—and so is sometimes known as *Equus Primus*—is Hippe: the daughter in hiding.

Hippe was the daughter of **Chiron**, the wise, kind centaur who raised **Asclepius** and **Achilles**, who saved **Prometheus**, and who shines in the night sky as Centaurus. Hippe loved her father very much, but he was sometimes very strict. She had worked very hard to learn all the words and numbers and musical notes he had taught her, and she knew more things than most girls, but she couldn't help feeling that still he wished she was a boy. One day when he was out teaching one of his many pupils to hunt—she, of course, had not been allowed to go— Hippe set out on a solitary walk. It was cold and the winds were up, but she was determined to have some adventure for once. So, wrapping herself up in a thick wool cloak, she made for the hills.

Perhaps it was she who saw Aeolus first; perhaps she was seduced willingly, perhaps she was not—legend does not seem to care. Either way, Hippe was soon pregnant. Full of shame and terrified that her father would find her, she hid in the mountains and begged the gods for help. At last, **Artemis** came to her rescue and turned her into a mare. Soon after, her foal, Melanippe, was born. **Poseidon**—who wanted to help his friend Aeolus—turned the little black horse into a girl and gave her to her father.

Artemis then placed Hippe in the stars, where she still hides behind Pegasus, with only her head peeping out.

CYGNUS

VULPECULA

SAGITTA

PEGASUS

DELPHINUS

AQUARIUS

AQUILA

CAPRICORNUS

0 1 2 3 4 5

MAGNITUDE

ERIDANUS

ERI/ERIDANI, THE RIVER

RANK IN SIZE: **6**
ASTERISMS: **NONE**

THROUGH ALL his boyhood, **Phaethon** dreamed of riding his father's chariot across the sky. He and his mother, Clymene, lived alone, in a shabby house at the edge of the town, and while the other boys knew that his mother was a sea nymph, none of them believed that his father could really be the great sun god **Helios**. Phaethon's bedroom walls were plastered with drawings of his father's chariot: he had spent boxes and boxes of yellow crayons tracing its journey through the skies; and when dawn arrived each morning he would leap out of his little bed (casting aside his Hercules-patterned sheets) and rush to the window to see if he could catch a glimpse of his father's broadening grin, lighting up the day. So when finally he had passed the last of his endless examinations, and his mother had sent him off into the world with a new pair of boots and a stifled sob, he headed straight for the steep, steep path that led to the gold and glimmering Palace of the Sun.

"What brings you here, my son?" asked the blazing Helios when the boy had reached its dazzling heights.

"Father"—he hesitated, shy—"no one believes that I am truly your son. I think if you would let me ride your chariot across the sky—just for one day—then I might show stupid Quintus, Flavius, and **Cycnus** once and for all."

Helios consented with the heaviest of hearts. For well he knew that the boy would be unable to control the fiery horses whose nostrils flared with anticipation as his son took up their golden reins.

He could hardly watch as Phaethon collided with constellation after constellation: first the fiery chariot brushed against the Little Bear and the Great Bear and set them both alight; then a ball of fire kicked from a horse's foot caught the end of Draco's tail, and the fearsome dragon, which had been snoring in his cave for centuries, was set terrifyingly ablaze. By the time that poor Phaethon saw Scorpius's vast pincers closing in toward him, he had lost his last shred of courage—and the reins, the horses, and the chariot to the sky.

The chariot plunged so low that it singed the land: Libya became a desert; the Ethiopians' skin was blackened; and the seas and rivers dried up. The Earth's wounded mother, **Gaia**, wailed until Almighty **Zeus** finally intervened. He hurled his most electric lightning bolt at the chariot, splintering it asunder and throwing the incensed horses to the sea.

And this is how Phaethon, his hair ablaze, leaving a trail of fire across the night, fell to his watery death in the depths of the river Eridanus.

FORNAX

FOR/FORNACIS, THE FURNACE

RANK IN SIZE: 41
ASTERISMS: NONE

THE ASTRONOMER Johann Bode's 1801 *Uranographia* reinvented this constellation to honor the French chemist **Antoine Lavoisier**. Bode fashioned this celestial furnace in the image of one of Lavoisier's seminal experiments: decomposing water into its components hydrogen and oxygen.

But Antoine Lavoisier was just thirteen when Fornax first found its way onto the astronomical map. When Nicolas Louis de Lacaille (or **Lacaille** as he is now prosaically known) sailed to South Africa to chart the southern stars, Lavoisier was still a gangly schoolboy at the Collège des Quatre-Nations in Paris, where he had not yet started the philosophy lessons he would go on to have there with the venerable Lacaille; nor had he been introduced to the wonders of meteorological observation by that same astronomer, let alone become, as he one day would, the father of modern chemistry.

So young Lavoisier may well have been staring anxiously at his upper lip in the glass—or a young mademoiselle on the boulevard—as Lacaille, sitting in his observatory on the shore of Table Bay, wondered quite how to dedicate another of his new constellations to the glories of science.

Of all the many and wondrous scientific instruments that had revolutionized the world, which should he choose? For it was 1751 (or perhaps 1752—we will never know exactly when it was that Fornax came into his view under the South African sky) and the Enlightenment was blasting with fury.

Aha! He had it. Blasting with fury indeed. He would depict the constellation as a chemist's furnace, joining the dots of the stars to show the distillation process: he would sketch a flask with a long neck being heated over a fire and another receptacle collecting the residue.

Returning from the Cape of Good Hope, Lacaille drew just this on his 1756 *Planisphere*, and, ever the Frenchman, labeled the constellation le *Forneau*. When a second edition was published in 1763, Lacaille caved in to astronomical convention and Latinized his furnace to *Fornax Chimiae* (the "Chemical Furnace").

Thirty-eight years later, however, the French revolutionaries had no such desire to honor the advances of science. In 1794, Lavoisier—who was all grown up and had helped develop the metric system, begun the first list of elements, and given both oxygen and hydrogen their names—was accused, among other things, of adulterating France's tobacco with water and marched off to the guillotine.

"*La République n'a pas besoin de savants ni de chimistes!*" (The Republic has no need for scientists nor chemists) they declared, before chopping off his head.

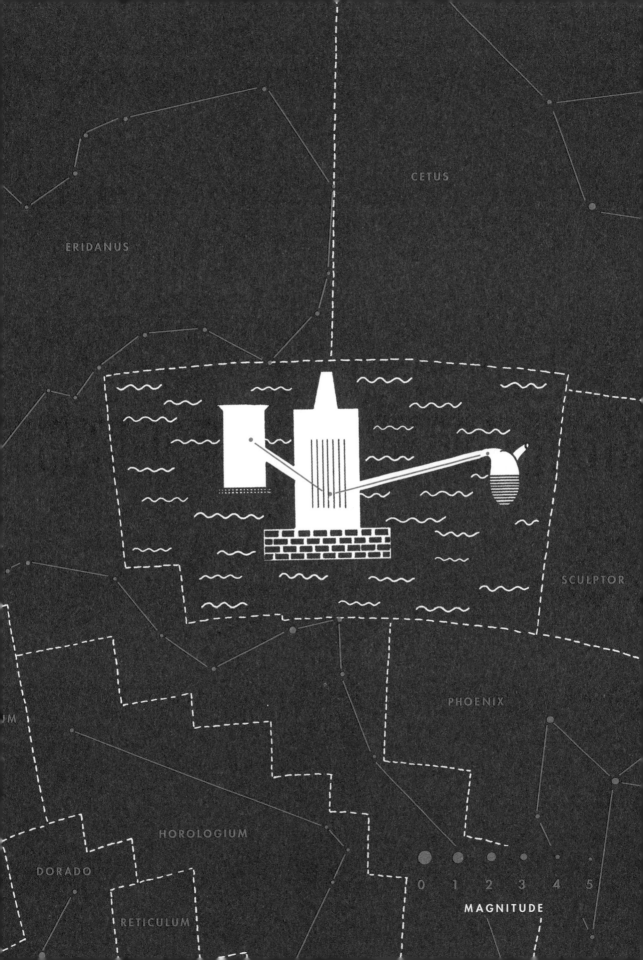

ERIDANUS

SCULPTOR

PHOENIX

HOROLOGIUM

DORADO

RETICULUM

0   1   2   3   4   5

MAGNITUDE

GEMINI
GEM/GEMINORUM, THE TWINS

TWO CHILDREN who grow in one womb, sometimes from one egg, sometimes from two: twins are a mysterious thing. Their duality makes them a symbol of both similarity and difference, of unity and of division. They are simultaneously stronger than most of us—they can share strange powers, secret languages, a sixth sense—and yet more vulnerable, especially apart. An anthropological curiosity of age-old fascination, inspiring wonder and awe, they were tested on obscenely in Nazi concentration camps; in Igbo-Ora—a small Nigerian town whose women inexplicably give birth to four times more twins than anywhere on Earth—they are worshipped as a special gift from God.

The two bright stars of Gemini are known as twins the world over. In Maori legend they are the mortal sons of Bora Bora. These two twin boys were devoted to each other: as thick as thieves in their private universe. But their parents worried that they didn't play with the other children, and like so many unfortunate parents, in trying to do the right thing, they ended up doing the worst. When the boys overheard that their parents were planning to split them up, they sneaked out of their beds in the middle of the night, stole into their father's boat, and sailed into the darkness. Awoken by maternal intuition, their mother saw their empty beds and ran to the shore. In the distance, their sail gleamed in the moonlight and she jumped into a neighbor's boat in pursuit. She chased them around island after island and all the way up a mountain in Tahiti, but just as she was about to reach her two little boys, they leapt off the top of the mountain and into the sky, where they remain forever as stars.

To the Greeks these lights were **Castor** and **Pollux**, one half of a pair of twins born to **Leda**, Queen of Sparta, in one of the most biologically and mythologically curious of all twin tales. After Leda's belly was filled by her rape by **Zeus**, who attacked her in the form of a swan—a horror remembered in the stars of Cygnus—she went home to her husband, **Tyndareus**, and, though she was bruised and torn, she let him satisfy himself too. The months passed. Some say she laid one egg, some say she laid two, but out of them hatched not just Castor and Pollux, the horseman and the boxer who joined **Jason** and the Argonauts in search of the Golden Fleece, but two of the women most blamed for and spattered in ancient blood, Helen and Clytemnestra.

What these twin stars were known as at one point in ancient China says it all: Yin and Yang.

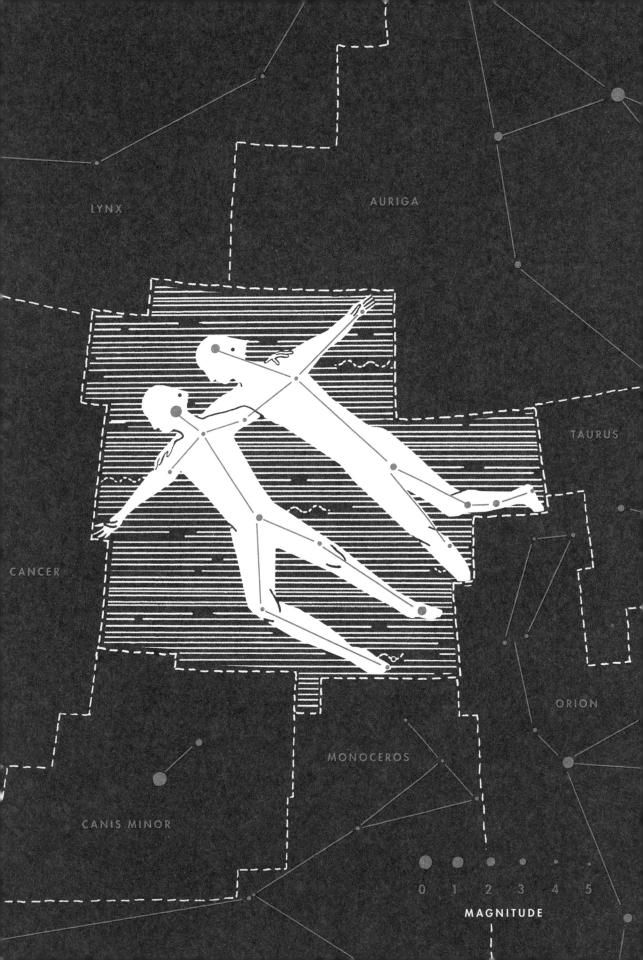

GRUS
GRU/GRUIS, THE CRANE

RANK IN SIZE: **45**
ASTERISMS: **NONE**

SADAKO SASAKI was two years old when she was hurled out of a window by a force hitherto unknown to mankind. She was born in Hiroshima in 1943, the first daughter of a family who owned a barbershop. Her father was drafted into the army while she was still a baby, and when an atom bomb exploded less than a mile away from the Sasaki family home, propelling their child into the street, her mother was sure she was dead. In fact Sadako had survived the blast without injury or burns, and her mother was able to pick her up and run through the black rain to safety.

Ten years later, Sadako had grown into a girl like any other at Nobori-cho Elementary School—except that she was by far the fastest runner on sports day—but she had started to notice strange lumps on her neck. The swellings didn't go away and soon purple spots appeared on her legs. She was diagnosed with leukemia and given less than a year to live. An ancient Japanese legend promises that a wish will be granted to anyone who makes a thousand origami cranes, and so Sadako set to work making hundreds and hundreds of colored birds, stringing them together and tying them above her hospital bed. When she ran out of paper she used medicine wrappers. Despite her folding more than a thousand cranes her wish never came true. She died on October 25, 1955.

The astronomer **Johann Bayer** cannot of course have been thinking of Sadako's story when he decided to depict the constellation discovered by **de Houtman** and **Keyser** in 1596 as a crane, instead of as a heron, as preferred by de Houtman; or a Phoenicopterus (a flamingo) as favored by **Petrus Plancius** and Pieter van den Keere; or indeed as a fishing rod as the people of the Marshall Islands in the Western Pacific saw Grus to be. He may, however, have been thinking about the biblical Stork of Heaven in Jeremiah VIII, with which the crane is associated:

6 *I hearkened and heard, but they spake not aright: no man repented him of his wickedness, saying, What have I done? Every one turned to his course, as the horse rusheth into the battle. 7 Yea, the stork in the heaven knoweth her appointed times; and the turtle and the crane and the swallow observe the time of their coming; but my people know not the judgment of the LORD . . . For they have healed the hurt of the daughter of my people slightly, saying, Peace, peace; when there is no peace. 12 Were they ashamed when they had committed abomination? Nay, they were not at all ashamed, neither could they blush . . .*

AQUARIUS

CAPRICORNUS

SCULPTOR

PISCIS AUSTRINUS

MICROSCOPIUM

PHOENIX

INDUS

TELESCOPIUM

TUCANA

0 1 2 3 4 5

MAGNITUDE

HERCULES
HER/HERCULIS, THE KNEELING HERO

....

RANK IN SIZE: *5*

ASTERISMS: **THE BUTTERFLY, THE KEYSTONE**

....

TO-DO LIST

1 Kill Nemean lion (N.B. has a hide impervious to any weapons—will have to
strangle it) ~~10 days??~~ *TOOK 30 DAYS—move schedule back*

2 Slaughter many-headed water-snake Hydra (lair is in swamp near town of Lerna
in the Peloponnese) ~~10 days??~~ *TOOK 30 DAYS—move schedule back*

3 Hunt down Ceryneian Hind (Diana's sacred pet) and bring back alive to King
Eurystheus in Mycenae ~~1 month??~~ *TOOK 1 YEAR—revise schedule entirely*

4 Capture Erymanthian boar and also bring back to King E alive
Ferocious—get in touch with my old teacher **Chiron** for advice on how to subdue it

5 Clean dung from Augean Stables ** IN 1 DAY **
Careful not to get lion pelt filthy—v. expensive to dry-clean

6 Rid Stymphalian marshes of flock of hideous birds
~~N.B. Test out if arrows still poisonous (after dipped them in Hydra's blood)~~
*DEFINITELY still working (. . .) MUST send Chiron a get-well-soon card (his knee is in
constant agony)—check he knows I did it by accident*

7 Go to Crete and deal with a white bull (??)—something about **King Minos** not
sacrificing it to **Neptune**/his wife giving birth to its son (the **Minotaur**)

8 Capture flesh-eating horses of **King Diomedes** of Thrace

9 Snatch (Queen of the Amazons) Hippolyte's snazzily decorated girdle for
King E's daughter, Admete

10 Steal cattle of Geryon (three-headed monster in land of Erytheia—far west,
beyond river of Ocean in place of the setting sun)

ADDITIONAL 2 TASKS JUST SET BY KING E
(The liar has completely gone back on his word—said there would only be 10 labors)

11 Fetch golden apples from garden of Hesperides on Mount Atlas
Potential problems:
  a) Secret location—no one knows how to get there
  b) Giant dragon called **Ladon** wrapped around tree protecting them (use club??)

12 Bring back **Cerberus** from the underworld
HOW???? (Three-headed guard dog with the tail of a dragon and a back covered
in snakes that devours anyone who tries to escape from the gates of hell)

URGENT: Buy life insurance

GNUS

DRACO

BOÖTES

LYRA

CORONA
BOREALIS

OPHIUCHUS
& SERPENS

0 1 2 3 4 5

MAGNITUDE

HOROLOGIUM
HOR/HOROLOGII, THE PENDULUM CLOCK

....

RANK IN SIZE: **58**
ASTERISMS: **NONE**

....

AT 5:20 A.M. on Friday, May 30, 2014, civilization as Britain knew it came to a terrifying standstill. For the first time in more than ninety years, BBC Radio 4 failed to broadcast the *Shipping Forecast*. Postmen didn't know if they should leave the house. Bleary-eyed commuters choked on their tea. A technical error robbed a bewildered public of this national stalwart, denying sailors and landlubbers alike the reassurance of the venerable program that had, until that point, been aired religiously four times a day since 1924. The horror! It's not just seafarers who wake to its calm tones and are lulled to sleep by its lilting waltz.

The histories of navigation and timekeeping are of course inextricably linked. This constellation, one of the fourteen created by **Lacaille** in 1751–2, was intended to commemorate the pendulum clock invented by Dutch scientist Christiaan Huygens in the 1650s. Originally called Horologium Oscillatorium, its alpha star marks the bottom of the pendulum. But a pendulum clock, as you might imagine, does not work at its best far out at sea. A problem that was not just serious, but fatal, for history's wayfarers. Right up until the eighteenth century, while captains could navigate their ships in straight lines—determining their route from the movement of the stars—they couldn't sail around corners. That is to say they could calculate their latitude, but not their longitude. Countless crews were pulverized against the rocks when they inadvertently aimed their ships directly at them; in 1707, an entire British naval fleet sank, and more than 1,400 sailors drowned, off the Isles of Scilly. As much as the celebrated explorers of the sixteenth century were skilled, they were also lucky: for every **Magellan**, Drake, and **Columbus** who made it back, there were several more who didn't. It seemed that not all aspiring colonialists had the Christian God they hoped to impress upon the natives on their side.

Eventually, the British government decided that something must be done. In 1714, it established the Board of Longitude and announced several prizes of up to £20,000 for a "Practicable and Useful" method that could determine longitude within sixty nautical miles. Dava Sobel's book, *Longitude: The True Story of a Lone Genius Who Solved the Greatest Scientific Problem of His Time*, tells the story of the man who dedicated his life to doing just that. In 1735, the Yorkshire carpenter John Harrison achieved what even **Newton** had thought was impossible: he designed the first marine chronometer accurate enough to measure the time difference between a ship's position and the Greenwich Meridian to therefore determine its longitude. In 1880, Greenwich Mean Time became the legal standard for the whole of Britain, and by 1920 the entire globe was setting its clocks to the pips first broadcast by the BBC in 1922.

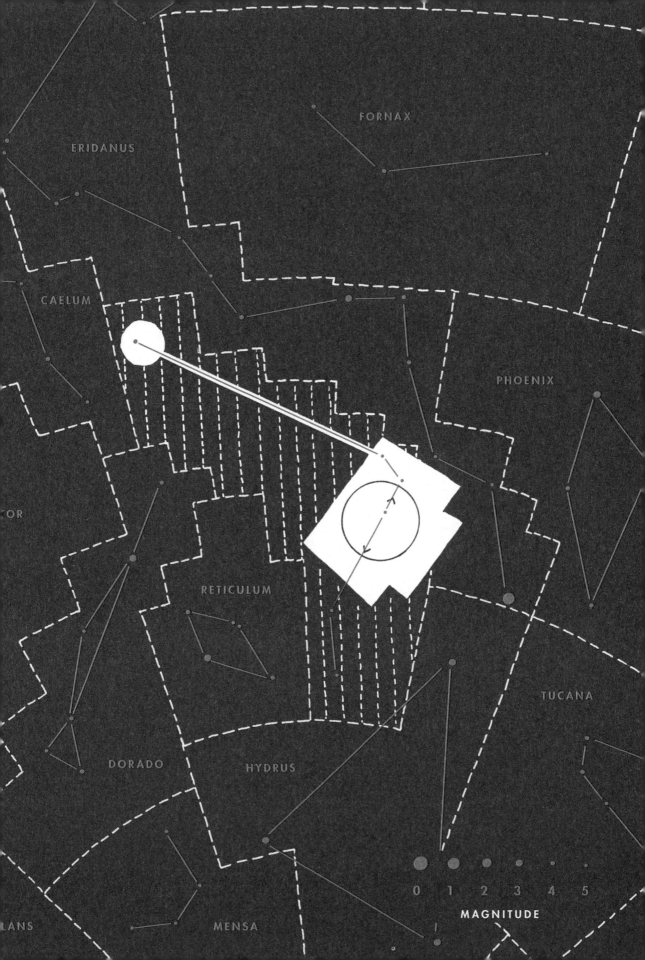

FORNAX

ERIDANUS

CAELUM

PHOENIX

OR

RETICULUM

TUCANA

DORADO

HYDRUS

MAGNITUDE

0   1   2   3   4   5

LANS

MENSA

HYDRA

HYA/HYDRAE, THE WATER-SNAKE

··

RANK IN SIZE: 1
ASTERISMS: **THE HEAD**

··

EXT. PELOPONNESIAN COUNTRYSIDE—DAY.

WIDE SHOT of the plains of the Argolid. We see cattle grazing and peasants at work in the fields.

CLOSE UP on a **YOUNG BOY** (Felix, 8) and his **SISTER** (Aemilia, 6), freckled, with dirty faces, helping to pick olives on their parents' farm. They are messing around, giggling. **PAN** to their **MOTHER** (20s) nearby in the olive groves. Once beautiful, but aged more than she should be, she is hard at work.

She stops dead in her tracks. She smells something.

CUT to:
WIDE SHOT of the foul and eerily still **LERNEAN SWAMP**.

CUT BACK to:
OLIVE GROVES. Camera zooms in on the **MOTHER**. She knows what's happening.

MOTHER
Felix! Aemilia!
(Her children are ignoring her cries. They are chasing each other around a barrel.)

MOTHER
(With increasing panic)
Felix! Felix! Get your sister inside.

CUT to:
LERNEAN SWAMP. TIGHTER SHOT. Something is moving under the fetid mud.

CUT to:
AEMILIA has now smelled the foul stench. She drops her basket of olives in terror and turns around to look for her brother. He is nowhere to be seen.

AEMILIA
(She knows what the smell signals.)
Felix! Felix!

CUT to:
LERNEAN SWAMP. The HYDRA (a vast serpent with nine heads and the monstrous
body of a scaled, reptilian dog) is emerging from its lair. We see it breathe out its
poisonous fumes.

CUT to:
FELIX, still behind the barrel where he was playing hide-and-seek, sees a huge shadow
loom over him, cutting out the golden Mediterranean light. He turns around to see the
gaping jaws of all the HYDRA's nine heads salivating and grimacing in his direction.

CUT to:
AERIAL shot over the scene. HYDRA'S P.O.V. Suddenly, as if from nowhere,
flaming arrows are being shot up at the monster.

CUT to:
It's HERCULES (long blond hair, ruggedly handsome, wearing a lion's pelt—its jaws at
his shoulder) in his chariot! He starts to attack HYDRA, swinging his club at the monster's
heads. But each time he destroys one, two grow back in its place. A vile crab (CANCER)
emerges from the swamp and attacks his toe, but heroic HERCULES just squashes it
underfoot while continuing his fierce battle with the HYDRA.

HERCULES
(Impressively calm and laconic amid the slaughter)
Iolaus. I think I'm going to need your help.

HERCULES's charioteer IOLAUS (late 30s, bearded, a loyal servant) dismounts from his
chariot—tying the wildly bucking horses to a tree. Joining HERCULES in the terrifying fray,
he sets fire to a branch using the flames of the HYDRA's breath. As HERCULES smashes
the monster's heads, he burns their stumps so they can't grow back.

Finally, HERCULES reaches the HYDRA's last, immortal head. IOLAUS hands him
ATHENE's precious sword, and he slices it off. He buries the head, still alive and writhing,
under a rock, and then slashes open the monster's body. As he dips his arrows in the
poisonous blood that is seeping out onto the earth, IOLAUS unties the horses. They mount
the chariot and, whisking up the amazed FELIX, gallop off into the distance.

LEO

CANCER

SEXTANS

CRATER

PUPPIS

PYXIS

ANTLIA

VELA

0 1 2 3 4 5

MAGNITUDE

HYDRUS

HYI/HYDRI, THE LESSER WATER-SNAKE

....

RANK IN SIZE: **61**
ASTERISMS: **NONE**

....

TWO JOKES ABOUT THE UNDISTINGUISHED
MALE WATER-SNAKE HYDRUS

(Not to be confused with the impressive and legendary female water-snake, Hydra, able to be slain only by the greatest of heroes, Hercules.)

• • •

How many lesser water-snakes
does it take to change a lightbulb?
Hydrustle up a few.

• • •

Knock, knock.

Who's there?

The Lesser Water-Snake.

The Lesser Water-Snake who?

The lesser water-snake does of
any celestial consequence,
the less there is to say about it.

• • •

(There are no myths associated with this constellation.)

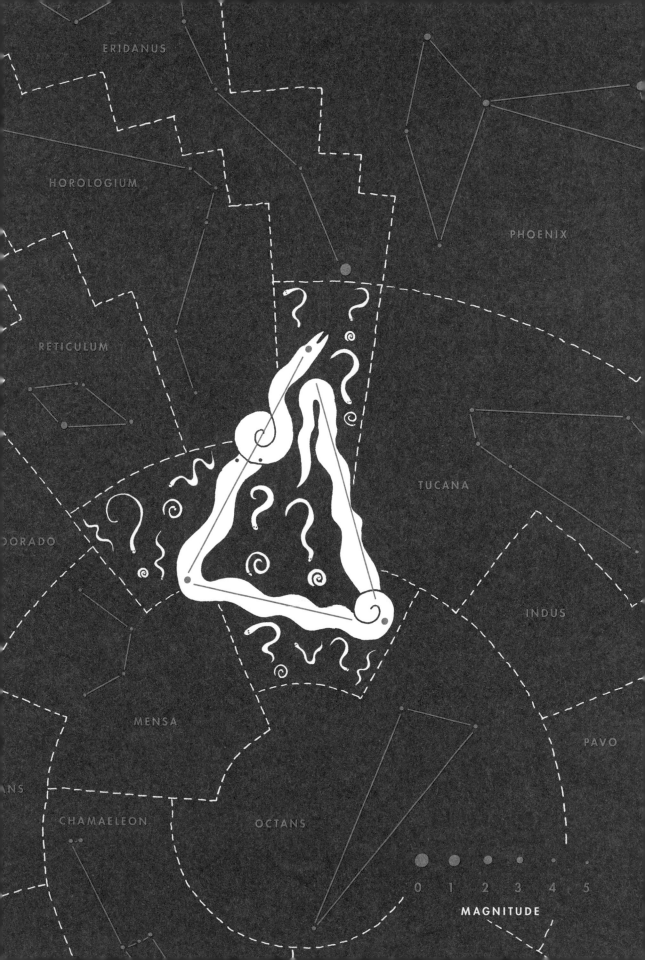

ERIDANUS

HOROLOGIUM

PHOENIX

RETICULUM

TUCANA

DORADO

INDUS

MENSA

PAVO

ANS

CHAMAELEON

OCTANS

● ● ● ● · ·
0 1 2 3 4 5

MAGNITUDE

INDUS

IND/INDI, THE INDIAN

RANK IN SIZE: **49**

ASTERISMS: **NONE**

THE ETHNICITY of the Indian in the sky is as obscure as its stars. At the end of the sixteenth century, when the Dutch navigators **Keyser** and **de Houtman** first formed this faint constellation, the word Indus (meaning "Indian") would have been used very loosely to describe the indigenous people of North or South America, as well as of the Indian subcontinent. But it is almost certain that these explorers were charting its stars from Madagascar, a long way from either continent, off the southeast coast of Africa. For it was there that the entire fleet of their voyage to the East Indies was forced to spend several months in 1595–6 to repair themselves and their ships; and it was from there that they made most of their astronomical observations.

This ambiguity reflects an unappealing truth about the mind-set of the time. Like the image depicting this constellation on contemporary star atlases—a grass-skirted or loinclothed tribesman wielding a spear—there didn't seem to have been a need to be particularly specific. In the ethnocentric European imagination at the turn of the seventeenth century, a half-naked Indian could stand symbolically for more or less any native of more or less any of the exotic places that explorers had only so recently discovered. On an even sourer note, it is telling that all the other figures that Keyser and de Houtman placed in the sky—Apus, Chamaeleon, Dorado, Volans, Pavo, Grus, Phoenix, Tucana, Hydrus, and Musca—were wild and exotic birds and beasts. It is hard not to draw the ugly conclusion that these early colonizers saw the people they came across in the same light.

That is not to say that Westerners weren't interested in these indigenous peoples. On the contrary, they were mesmerized—and disturbed—by these new encounters, as both Michel de Montaigne's vision of the "noble savage" and Shakespeare's *The Tempest* attest. Written in 1610 only a few years after Indus was created (the constellation first appeared in 1598 on a globe by **Petrus Plancius**, and first appeared in print in **Johann Bayer**'s 1603 *Uranometria*), Shakespeare's play centers on the thorny relationship between its protagonist, Prospero, and two natives of the remote island he is shipwrecked on. Ariel and Caliban, whose very names allude to myth, magic, and cannibalism, reveal the conflicting feelings that Jacobean England had to the alien inhabitants of strange lands: fascination and fear in equal measure.

PISCIS AUSTRINUS

MICROSCOPIUM

GRUS

SAGITTARIUS

TELESCOPIUM

NIX

PAVO

TUCANA

OCTANS

APUS

HYDRUS

LACERTA

LAC/LACERTAE, THE LIZARD

....

RANK IN SIZE: **68**

ASTERISMS: **NONE**

....

I AM TRYING to catch a lizard, but it keeps slipping from my grasp. I have chased it across the Internet, but every time I have it in my sights, it slithers away into a cyber crack. I have looked for it in books, but it's always hiding somewhere in between the lines.

I first caught sight of the squirming little reptile in the frontispiece of the Polish astronomer **Johannes Hevelius**'s star atlas, his *Firmamentum Sobiescianum*, dated 1687 but not published until after his death in 1690. The opening pages of this exquisitely illustrated book feature an engraving of Hevelius presenting his new constellations to **Urania**, the muse of astronomy. An extraordinary image, it depicts the (not so convincingly) humble stargazer genuflecting in front of a globe, holding a shield and an astronomical instrument (to represent two of the new constellations, Scutum and Sextans) and looking up to his muse, who, beaming and brilliant, is holding the Sun and the Moon in her hands, and is flanked on all sides by the great heroes of astronomy: **Hipparchus**, **Brahe**, **Ptolemy**, and **Copernicus** among others. Behind their benefactor trot a procession of the animals that he has placed in the sky—a fox and the goose (Vulpecula), two hunting dogs (Canes Venatici), a lion cub (Leo Minor), and a Lynx—all being led by none other than our slippery friend, Lacerta. Yet for all the pomp and glory of his star atlas's frontispiece, Hevelius doesn't seem to have been much interested in this little lizard. Indeed he seems to have been so unconcerned about the particulars of this astral creature that he even offered an alternative moniker for it: Stellio ("the stellion"), the name of a Mediterranean newt, which did at least have starlike markings on its rough back. So almost as quickly as I had clapped eyes on the lizard, it escaped from my grasp.

Next I saw it lurking in the corner of a Navajo rug. It was hiding in its geometrical patterns, mirroring the shapes of the string games that Navajo medicine men teach their tribe's children to show them the constellations that the First Man arranged in the sky. But it was hard to know if this was the same lizard that I was chasing. For the very last time I caught sight of it, it had morphed shape almost out of all recognition and was now a flying serpent called Tengshe, the Chinese asterism that holds northern Lacerta in its center.

It turns out that the lizard secreted between the stars will confound you just as frustratingly as its earthly counterparts do between the rocks.

DRACO

CEPHEUS

CASSIOPEIA

CYGNUS

ANDROMEDA

EQUULEUS

PEGASUS

0 1 2 3 4 5

MAGNITUDE

LEO

LEO/LEONIS, THE LION

....

RANK IN SIZE: **12**

ASTERISMS: **THE DIAMOND (OF VIRGO), THE SICKLE, THE SPRING TRIANGLE**

....

HE WAS BORN to play Hercules, he told himself encouragingly as he walked through a rainy Soho, stopping occasionally to check his hair in the glass of a coffee-shop window. He rang the third-floor buzzer. He was pretty sure it was a 3 that was peeling off the fading typeface of *Starlight Casting Suite*. Walking into the waiting room he saw a line of men in their twenties, arranged on cheap chairs as if they were variations on a parody—of himself. Some of their hair was curly, some was straight, but all of it was blond and fell to their muscly shoulders, just like his. He noticed with derisive amusement that one of them was wearing Ugg boots. But then wondered just as quickly—and in mild alarm—if this intimation of wilderness actually gave that particular doppelgänger an edge.

In between giggles down the telephone, the girl at the desk told him they were running forty minutes behind schedule. He sat down on a lime green plastic seat in between a skinnier version of himself on his left, and a younger one on his right. He flicked idly through *Metro*, stopping only to read an article about a new caveman diet, and his star sign. Finally his name was called.

A balding man with ludicrously colored sneakers was talking to him in a nasal drone:

"So you're in a cave, and there are loads of hot girls around—'cuz that's what this lion does, you know. Drags all the town's women into his lair. And they're all looking up at you, like, pleading and terrified—but still hot." The man laughed unattractively. "We see you block out one of the entrances to the cave, and then we cut to you grabbing the lion in your arms—and make sure you give us a really good shot of your biceps, yeah—and then you kill it, save the girls, and do the tagline. OK? Action."

"Sorry, but, um, do you want me to mime the club?"

"Nah, mate, the whole point is weapons don't work on this lion—it's got this impenetrable golden fur—we're doing this really jazzy thing with it in CGI. You strangle it with your bare hands."

As he threw himself unconvincingly around a square foot of blue carpet, throttling the sterile air for the camera, he wondered if he had made the best decisions he could in life. But then he thought of himself, bronzed and topless, with a lion's pelt about his shoulders; and he thought of his burgeoning overdraft and his leaking radiator; and Leo Clark hurled himself into Hercules's First Labor with fury.

LYNX

URSA MAJOR

LEO MINOR

SEXTANS

HYDRA

GO

CRATER

0 1 2 3 4 5

MAGNITUDE

LEO MINOR

LMI/LEONIS MINORIS, THE LION CUB

RANK IN SIZE: **64**
ASTERISMS: **NONE**

A TRUE STORY ABOUT AN INSIGNIFICANT
NORTHERN CONSTELLATION

SHE HAD FALLEN in love with astronomy before she could remember. When still a little girl, **Catherina Elisabetha Koopman** crossed the streets of her native Danzig with her parents (rich merchants able to fulfill their daughter's fantasies) and knocked on an astronomer's door. Showing her the delights of his world-famous observatory, **Johannes Hevelius** promised to unveil to her—when, of course, she was old enough to understand—the mysteries of the night sky.

Elisabetha passed through the years with impatience. Finally, when she was fifteen, she knocked on his door again. Thirty-six years her senior, this proud lion of the astronomical world gave a broad grin. Hevelius's first wife, who had not shared his passion for the stars (but preferred to nag him about the finances of the family brewery he was forced to run), had just died. We will never know at what exact point after Elisabetha stepped over his threshold the consummate mathematician put two and two together, but years later he recalled their union with great sentiment:

> When in the star-lit night she followed with enraptured gaze and beating heart, through his giant telescope, the shining full moon, on her silent path, she exclaimed with enthusiasm, "To remain and gaze here always, to be allowed to explore and proclaim with you the wonder of the heavens; that would make me perfectly happy!" And the worthy man felt that it might make him happy too.

His "loyal helpmate" may not have seen it quite so romantically. As a woman, she would not be allowed to attend university, and the only way to pursue a scientific career would be through an expedient marriage. Perhaps in Elisabetha's case, however, opportunity found itself intermingled with passion—a Polish Dorothea, whose heart jumped at this aging man's erudition and longed, with an ardor rivaling even that of George Eliot's zealous heroine, to commit her youth and beauty to his magnum opus. And for his part, Hevelius was no unaffectionate tyrant: when his young wife fell ill with smallpox he never left her bedside.

Hevelius died before the star atlas that he had been working on (with the considerable aid of his wife) was finished. His *Firmamentum Sobiescianum*, edited and published by Elisabetha in 1690, includes the insignificant and often ignored Leo Minor. Mrs. Johannes Hevelius, whom François Arago called "the first woman, to my knowledge, who was not frightened to face the fatigue of making astronomical observations and calculations," is considered the first female astronomer. A minor planet, 12625 Koopman, has been named in her honor.

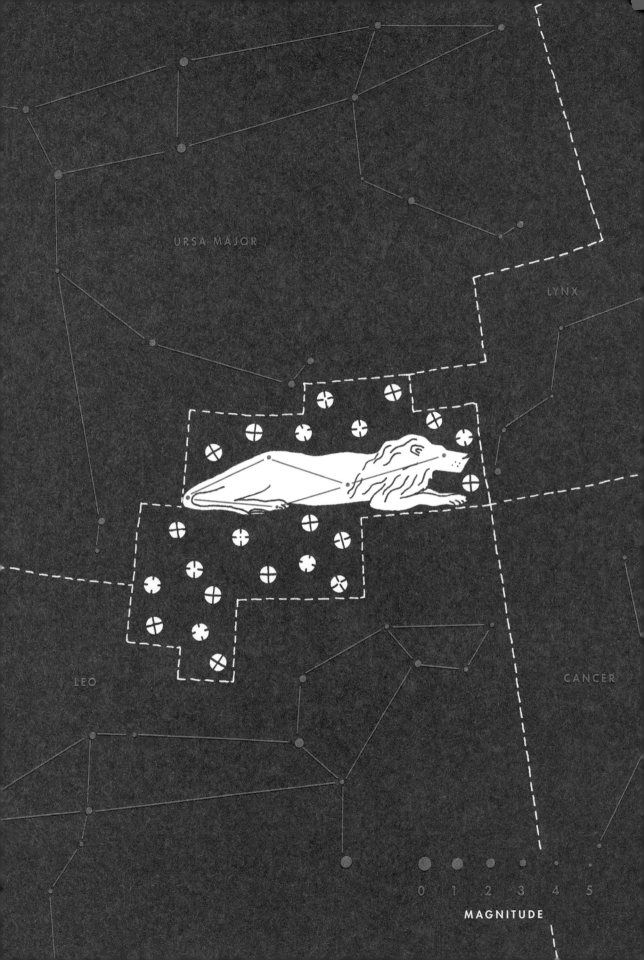

URSA MAJOR

LYNX

LEO

CANCER

0 1 2 3 4 5

MAGNITUDE

LEPUS

LEP/LEPORIS, THE HARE

RANK IN SIZE: 51
ASTERISMS: NONE

CROUCHING IN FEAR beneath the feet of the great hunter Orion, and hiding from his dogs—Canis Major and Canis Minor—as they try to sniff out its scent, is Lepus. This timid hare plays a bit part in the dramatic hunting scene of an action film that is projected onto the night sky of the northern hemisphere every winter. Largely forgotten by the other characters, and in the corner of the shot, the hare—about to be killed by Orion—is spared in the nick of time as Taurus the bull comes charging at the huntsman with its terrifying horns. Orion is not so lucky: as the stars of Scorpius rise in the east, the deadly scorpion rises up from a crack in the ground to deliver a fatal bite to the huntsman's foot, and the stars of Orion set, defeated, in the west. **Asclepius**, the great healer, then bounds into the shot and uses all the medical cunning taught to him by the wise **Chiron** (Centaurus) to revive him. As he slowly brings the huntsman back from the dead, Scorpius is once again crushed back down into the earth, and Orion rises again in the east.

It makes sense that when this celestial hunting film was first watched in 4000 BC, it would have been screened in the autumn—in hunting season—whereas the Earth's celestial precession over the years means that now it is projected most vividly in winter. The hare, as mentioned, had only a minor part, but he was so well reviewed that he is remembered in the sky just as well as the lead actors. A nineteenth-century ornithologist even bumped up his role: adding a layer of psychological complexity, D'Arcy Thompson created a new Hitchcockian spin to the drama. According to him, the hare is pathologically afraid of ravens, and so as Corvus the crow rises, poor Lepus scampers away in panic, burrowing deep into his dark warren beneath the sky.

Chinese astronomy also sees an annual hunt in this part of the sky, although one with some slightly less delicate imagery. In that ancient tradition, the stars Alpha, Beta, Gamma, and Delta Leporis form *Ce*—a toilet—while Mu and Epsilon Leporis create a privacy screen for hunters wanting to use this celestial bathroom. There is even a star assigned to depict their feces, dropping south from *Ce*, which lies in the constellation Western stargazers call Columba.

TAURUS

ORION

MONOCEROS

ERIDANUS

CANIS MAJOR

COLUMBA

CAELUM

PUPPIS

PICTOR

0 1 2 3 4 5

MAGNITUDE

LIBRA

LIB/LIBRAE, THE SCALES

RANK IN SIZE: **29**
ASTERISMS: **NONE**

ARE YOU A LIBRA? Balanced, just, and diplomatic? Born into the sign of the zodiac associated with the scales since ancient times? Perhaps because 2,000 years ago the Sun passing into Libra marked the September equinox, when day and night were equal.

In the medieval world, astrology was synonymous with astronomy and a sophisticated and highly respected science. A horoscope was not something you leafed to guiltily over your Special K but a meticulously drawn astrological chart revealing the political significance of planetary configurations observed to the degree and minute using elaborate instruments called astrolabes and quadrants.

Horoscopes may have their roots in personal predictions—ancient examples of which survive from fifth-century BC Babylonia and Egypt—but by the time the twelfth-century scholar Adelard of Bath introduced the first accurate astronomical tables to the west, horoscopes were a far more sophisticated affair. The Greeks had developed the ancient art, thrown in a bit of philosophy, and passed it on to the Arab world, where it assimilated Indian, Persian, and Islamic traditions. So that when medieval English scholars traveled to Spain, Sicily, and the Middle East, they brought home a vast body of scientific work including astrological, alchemical, and magical texts.

Of the five key planetary aspects (that is, angular relationships between celestial configurations), medieval astrologers saw particular significance in conjunction—the appearance of planets on the same degrees of longitude. Conjunction foretold of major historical happenings, of religious and political upheaval. So when it became apparent that in September 1186, all of the seven then known planets (the Moon, Mercury, Venus, the Sun, Mars, Jupiter, and Saturn) would conjoin in the sign of Libra, astrologers had a predictive field day.

The English chronicler Roger of Hoveden catalogued the panic. Since Libra was classified as an airy sign, there were widespread fears of violent storms and winds. Astrologers like Corumphiza warned of blackened air polluted "with the stink of poisonous vapours": "many people will be seized by death and sickness, and loud noises and voices will be heard in the air, terrifying the souls of those who are listening." As the day approached, alarm spread.

But luckily—and as so disappointingly with even the most hopeful of newspaper horoscopes—reality bore only a passing resemblance to prophecy. Nothing more impressive occurred than the eternally British scenario of some hail in Kent and some floods in Wales.

OPHIUCHUS
& SERPENS

LES

VIRGO

SCORPIUS

HYDRA

LUPUS

CENTAURUS

0 1 2 3 4 5

MAGNITUDE

LUPUS

LUP/LUPI, THE WOLF

RANK IN SIZE: **46**
ASTERISMS: **NONE**

THIS WILD ANIMAL has shape-shifted its way through time. To the Akkadians it was *Urbat*, a beast associated with death; to the Babylonians it was *Ur Idim*—a wild dog or wolf. The ancient Arabs saw it as *Al Asadha*, the lioness; the ancient Turks as an unspecified but ferocious being; so too the Greeks, who called it *Therium*; and the Romans *Bestia* or *Fera*. Sometimes the savage animal was a hybrid, with the head and torso of a human, and the legs and tail of a lion. The Greeks saw a wild creature impaled on a spear being carried by Centaurus to the nearby altar of Ara—possibly the raging Erymanthian boar that Hercules drove into thick snow, bound, and captured. But men have always tried to tame wild beasts with stories. When Renaissance scholars translated **Ptolemy** from Greek into Latin, they found a good Greek myth to make sense of this amorphous creature: to fix it finally in one shape.

...

King Lycaon had civilized Arcadia, building the first city in that remote and mountainous sylvan idyll. Lycaon was a cruel and arrogant king with fifty vicious sons, and his impiety was famous across the lands: good Grecians would shudder in horror to hear of the heresy of Lycaonian rites—barbarous and cannibalistic rituals performed in the name of **Zeus**. When talk of Lycaon's wicked crimes reached Mount Olympus, Zeus descended to Earth disguised as a poor laborer. Knocking on the door of Lycaon's great palace, Zeus was received with seeming *Xenia*—the sacred practice of hospitality and guest-friendship hallowed among the Greeks. But one may smile, and smile, and be a villain. The supper that Lycaon and his sons served the camouflaged god was a foul soup made from the umbles of sheep and goats and the flesh of Zeus's son **Arcas**. Recognizing at once the taste of his own son on his tongue, Zeus's fury knew no bounds: he overturned the table and struck thunderbolts into every corner of the lofty dining hall, scorching to death all fifty of Lycaon's sons. The king fled from the god's rage, but as he ran his screams turned to howls and his limbs grew hairs and yellow fangs pierced their way through his mouth. Zeus turned Lycaon—whose name means "of the she-wolf"—into a wolf that attacked the king's own sheep. But what did Zeus do with the remains of his butchered son? Look up to the constellation Boötes—the answer is in the stars.

LIBRA

VIRGO

IUCHUS
ERPENS

HYDRA

CENTAURUS

SCORPIUS

ARA

NORMA

CIRCINUS

TRIANGULUM
AUSTRALE

MAGNITUDE

0   1   2   3   4   5

MUSCA

LYNX

LYN/LYNCIS, THE LYNX

RANK IN SIZE: **67**
ASTERISMS: **NONE**

THE SEVENTEENTH-CENTURY Polish astronomer **Johannes Hevelius** was not the most modest of men. Especially when it came to his eyesight. Despite the universe-shifting discovery of the telescope around the time of his birth (he was born in Danzig in 1611, two years after **Galileo** had begun using one in Italy), he maintained that his own eyes were a far preferable instrument. Indeed he often exaggerated quite how faint some of the stars he catalogued actually were. He was not, of course, using his naked eye alone to make his observations: his addition of ten new constellations to the Greek astronomer **Ptolemy**'s original forty-eight was aided by astronomical instruments such as the sextant and the quadrant. Moreover, some of the stars in the seven of his constellations remaining on the celestial map were already known to the Greeks.

Nevertheless, in the case of Lynx, his ocular pride was well founded. You would indeed need the searing vision of that short-tailed wild cat in order to spot this faint cluster of stars, shining (with one exception) no brighter than fourth magnitude, in a large and bare expanse of sky between Auriga and Ursa Major, north of Gemini. It is interesting that Hevelius should have placed this luminous-eyed beast near those famous twins. For though he seems to have ascribed no legend—other than his own—to his placing this keen-sighted feline in the sky, there is a story that ties together the Greek twins and this Polish lynx.

One Hellenic day **Castor** and **Polydeuces**, the *Dioscuri* ("Boys of **Zeus**"), got into a squabble with their cousins, Idas and Lynceus, who were also twins. It was perhaps not wise of Castor and Polydeuces to steal their cousins' brides from them, particularly not on the wedding day itself; nor was it best for familial relations to whisk the beauties off to Sparta and then impregnate them. They must, surely, have expected revenge.

Pursuing his cousins after the ruined nuptials, Lynceus—who had vision so superhuman he could even see through solid objects—ran to the top of a mountain and, lynx-eyed, scanned the land below. Spying the boys crouching in the hollow of an oak tree, he beckoned his brother and they crept up on the treacherous pair. Stabbing into their hiding spot, Lynceus fatally impaled a startled Castor. Polydeuces—immortal and full of fury—drove his sword into Lynceus, while Zeus himself came to his son's aid, striking down Idas with a thunderbolt. Taking his dying brother, blood spurting from his mouth, in his arms, Polydeuces begged mighty Zeus to let him die alongside his beloved twin. Touched by his son's devotion to his mortal sibling, the god united them forever in the stars.

DRACO

CAMELOPARDALIS

URSA MAJOR

AURIGA

MINOR

GEMINI

CANCER

0 1 2 3 4 5

MAGNITUDE

LYRA
LYR/LYRAE, THE LYRE

....

RANK IN SIZE: **52**
ASTERISMS: **THE SUMMER TRIANGLE**

....

WE'RE LYING in bed trying to think of your name. He wants to call you Bob.

"Bob?! As in Bob your uncle, Bob the Builder?"

"As in Dylan, for Christ's sake. Bob Dylan. The greatest musician the world has ever known."

I tell him that I've always quite liked Gertie. He says you're not a girl; and that even if you are a girl, you're not some sort of saucy Victorian maid. I suggest Leonard. As in Cohen.

"Leonard is the boy in the corner of the playground who gets beaten up because he sounds like he's in the 1930s, isn't wearing sneakers, and has probably wet himself."

I turn the light off and roll over. Or rather I try instinctively to roll over but then roll back as my body—and you—remind me I can't. I don't get to sleep for hours, instead staring out of the window looking at the couple of stars I can see through the city's non-darkness. Imagine naming the stars. How on earth did anyone do that? Someone once told me about a website where you can buy a star and give it your own name. I try to remember what I learned about constellations in school. The Big Dipper, Orion's Belt, I vaguely remember something about Andromeda and **Ovid**'s *Metamorphoses*. Is there an **Orpheus**? Your father might go for Orpheus—the folksinger of all folksingers; the original bard; the poet who sang so beautifully that all the animals would stop and listen as his guitar gently wept. (Was it a guitar he played?) But then I think of what happened to Orpheus. First of all there was looking back at Eurydice in the underworld and losing her forever, and then he was torn apart, bone by bone, by a band of women in a bacchic frenzy. I think about how all the good stories and all the good names are full of pain, and I wonder how we could possibly bequeath you all that.

...

I'm lying in bed and you've just been born. You *are* a girl. You open your mouth and the sounds you cry are as beautiful to us as all the notes of all the best songs we have ever heard rolled together into a bawl. Lyra. That's what we call you, Lyra.

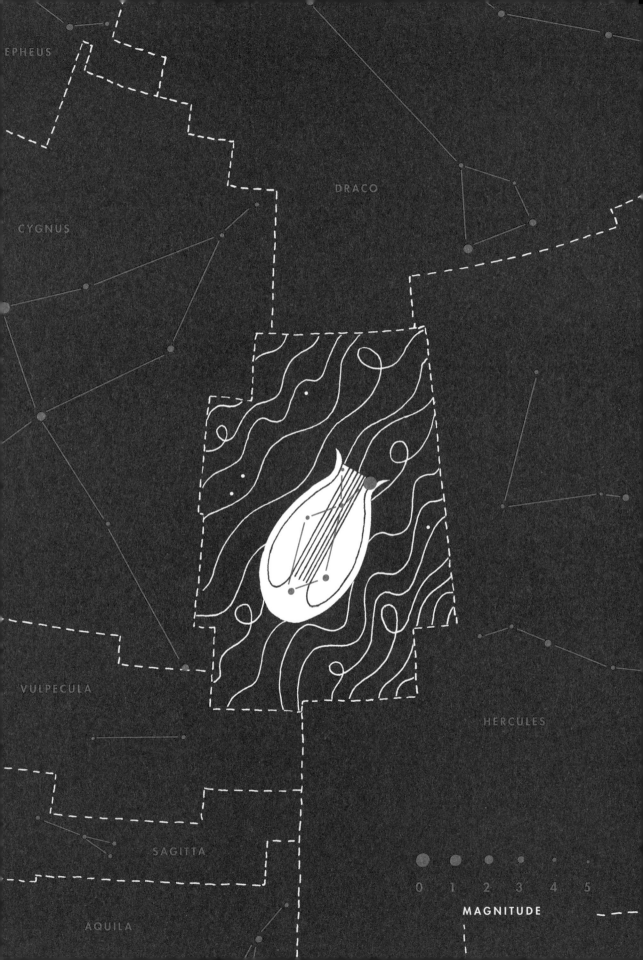

EPHEUS

CYGNUS

DRACO

VULPECULA

HERCULES

SAGITTA

AQUILA

0 1 2 3 4 5

MAGNITUDE

MENSA

MEN/MENSAE, THE TABLE MOUNTAIN

RANK IN SIZE: **75**

ASTERISMS: **NONE**

THE ONLY LANDMASS to make it into the heavens, Table Mountain in South Africa's Cape of Good Hope loomed over **Nicolas Louis de Lacaille** as he catalogued the southern stars in 1751–2. As he sat in the town below this flat-topped wonder, gazing at the dazzling vistas above, did he have any idea of the hundreds of ancient stories that haunt this magical place?

One such legend is the creation story that the Xhosa tell. The sun god Thixo and the one-eyed earth goddess Djobela had a son, Qamata, who was trying to create the world. Nganyamba, the Great Dragon of the Sea, was unsurprisingly a little riled by this young upstart who wanted to break up his primal oceans with some newfangled notion of dry land, and the two went head-to-head. Seeing her son struggling and injured in the battle, Djobela came to his aid. Summoning up four mighty giants, she placed one in each corner of the newly made Earth to guard it from the jealous dragon. But even they were not strong enough to resist Nganyamba's force. As each of them died, they asked Djobela to turn them into mountains so that they could continue to protect the land, even in death. And this is how Umlindi Wemingizimu, the watcher of the south and the greatest giant of all, became Table Mountain.

When Lacaille placed this table in the stars as Mensa, he envisioned the Large Magellanic Cloud, of which it contains a part, as its white tablecloth. This neatly mirrored the real cloud that often hangs over the iconic landmark, about which there is a Dutch legend.

Jan van Hunks was a man who liked to smoke. He lived at the foot of the mountain, and all day long he would sit on its slopes puffing away at his trusty pipe. When a passing stranger challenged him to a smoking competition, van Hunks let out a chesty laugh. Little did he know he had just been propositioned by the Devil himself. The two men smoked their way through a pile of tobacco until the air was so thick they could no longer see beyond the ends of their pipes. It did not surprise the Dutchman that the stranger caved in first, but when his competitor, lurching forward with nausea, dropped the hat from his head, he was astounded to see who he had beaten. Van Hunks had only a few seconds to revel in his victory before the irate Devil clapped his hands and vanished them into thin air. Their smoke, however, remained, hanging in a huge cloud over Table Mountain. The summit they had sat under was known ever after as the Devil's Peak.

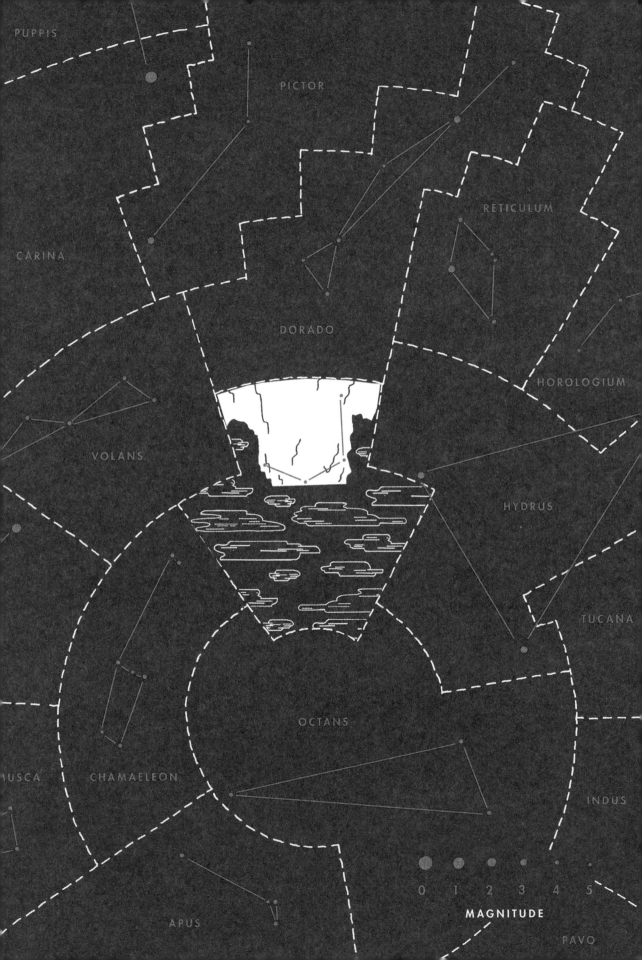

MICROSCOPIUM

MIC/MICROSCOPII, THE MICROSCOPE

RANK IN SIZE: **66**

ASTERISMS: **NONE**

ANOTHER ONE of the figures invented by the literal-minded **Lacaille** in the 1750s to commemorate the scientific achievement of his age of Enlightenment, Microscopium cuts as pedantic a figure in the sky as its creator does in the history books. Like many of the French astronomer's constellations, most of its fifth-magnitude stars are as invisible to the naked eye as the amoeba wriggling on the slide of its earthly counterpart.

Since we are already contemplating the hairsplitting of the firmament that this stellar instrument represents, it might be an apt moment to bring up the way in which the sky is officially dissected today. While archaeological findings have revealed that humans seem to have been mapping the stars for as many as 17,300 years—astronomical cave paintings discovered in Lascaux in southern France are thought to be representations of the *Pleiades* and the *Hyades*—it wasn't until 1928 that they got around to agreeing on an official system of how to do so.

Originally, constellations were very much just asterisms: that is, figures defined by the shapes they made in the stars, handed down through legend and lore. In AD 150, the ancient Greek astronomer **Ptolemy** catalogued the thousand or so stars then known to exist—the result was his seminal work that the Arabs later titled the *Almagest*—and arranged them into forty-eight constellations in this manner. Subsequent Western astronomers followed his lead, and any new stars discovered—whether by Dutch explorers in the sixteenth century or Polish, French, or German scientists in the eighteenth—were always arranged into constellations envisioned as specific patterns in the night sky (however little they actually resembled their eponymous forms). Interestingly, the astronomies of many other cultures, for example Aboriginal Australians, do not follow this "connect-the-dots" logic.

By the twentieth century, however, advances in celestial observation had outgrown the mythical system of more innocent days. There were so many stars in the modern sky that there was a need for a clearer system. Moreover, the lines that earlier cartographers had drawn on their atlases to demarcate the different asterisms were haphazard and inconsistent. In 1922, the first General Assembly of the International Astronomical Union adopted the list of the eighty-eight constellations that we use today. Six years later, the Belgian astronomer **Eugène Delporte** presented a definitive list of their boundaries to the IAU, and this carving up of the celestial globe has been conformed to ever since.

In the most official astronomical depictions of the sky, there are merely IAU boundary lines to map the heavens. The old animals, gods, and heroes have disappeared. And even Lacaille's seemingly unimaginative scientific instruments have disappeared into the ether.

AQUILA

AQUARIUS

CAPRICORNUS

PISCIS AUSTRINUS

SAGITTARIUS

GRUS

TELESCOPIUM

INDUS

0 1 2 3 4 5

MAGNITUDE

MONOCEROS

MON/MONOCEROTIS, THE UNICORN

RANK IN SIZE: **35**
ASTERISMS: **NONE**

"**THERE'S NO SUCH** things as unicorns, pea-brain," her brother said with a smirk. "Or Father Christmas. Unicorns are just something you get in *Alice's Adventures in Wonderland*, and Father Christmas was made up by capitalism." He pronounced the last word with particular pride.

Betty watched in disgust as her brother guzzled his Milky Way bar. He didn't even know his seven times tables, and whenever Mom tested him in the car, Dad secretly mouthed the answers to him in the rearview mirror. She thought with pleasure of the fact that soon he would have eaten all his chocolate and opened all the presents from his stocking, while her careful, steady pace meant that she had still unwrapped only one gift—a pinkly glittering My Little Pony. She would have lots left and he would be really annoyed.

"There is too such a thing as a unicorn," she said calmly. "They knew about them in ancient India and China, and you can see pictures of one fighting a lion from 3,500 years ago. And also medieval tapestries tell you how to hunt one. Which is basically that you get something called a virgin and then the unicorn, which is usually really shy, comes and sits in her lap and she strokes its head and it falls asleep and then they catch it. Which is all to do with something called paganism and f . . ."

Her brother, whose mouth had fallen agog mid–Milky Way some time ago, leapt to attention, sensing a moment to pounce.

"Fertility," she continued hurriedly, "and falling in love and magic. But that's before the Christians came along and stole the unicorn from the pagans and said it was actually called Jesus, and that the virgin woman was called Mary and it was all about something called the Passion. Which I think is a bit like the fruit. Except that in the medieval tapestry it's a pomegranate. And, poo-brain, the unicorn is in *Through the Looking-Glass* in a song about plum cake and a fight between Scotland and England that happened in real life, and that's why there's all those lions and unicorns in Buckingham Palace.

"And if you still don't believe in unicorns," his little sister said defiantly, "go and look through a telescope and you can see one—except it's called Monoceros because you have to call the animals in the stars by their Latin names. You can see its horn, which is a cone-shaped pillar of dust and gas 2,700 light-years away, which is even more magic than anyone ever imagined because it makes actual stars. And it made a beautiful pattern of them called the Christmas Tree star cluster.

"So, pea-brain, there are such things as unicorns and there probably is a Father Christmas too."

That shut him up.

TAURUS

ORION

ERIDANUS

CANIS MAJOR

LEPUS

MAGNITUDE

0 1 2 3 4 5

MUSCA

MUS/MUSCAE, THE FLY

···

RANK IN SIZE: *77*

ASTERISMS: **NONE**

···

YOU CAN'T CATCH A FLY. They get everywhere. Even into the nooks and crannies of outer space, when they were supposed in fact to be a bee.

This celestial fly was invented by the Dutch explorers **Keyser** and **de Houtman** in the late sixteenth century, using the southern hemisphere stars they observed on the first trading expeditions to the East Indies, then known as the *Eerste Schipvaart*. Keyser, who was the chief pilot on two of the four ships that left the Netherlands in 1595, the *Hollandia* and the *Mauritius*, had been instructed by the astronomer and cartographer **Petrus Plancius** to fill in the uncharted gaps in the sky around the South Celestial Pole. By the time the fleet reached Madagascar, almost a third of the sailors had died, mostly from scurvy. As the depleted crew spent several months anchored there, repairing the ship and themselves, Keyser, according to the logbook, "sought comfort in science." Standing in the crow's nest and using an astronomical instrument given to him by Plancius—probably either a cross-staff or an astrolabe—he "enriched his knowledge in astronomy by improving the position of old and the observation of new constellations."

Not much else is known about this intrepid navigator who died soon after at Bantam (now Banten, in Java) in 1596. But the stars he catalogued with the help of de Houtman were brought back to Plancius in Amsterdam, who illustrated them on his 1598 globe, leaving this particular constellation curiously unnamed. Despite the fact that de Houtman's catalogue listed it as *De Vlieghe* (the Dutch for "fly"), the German astronomer **Johann Bayer**, not realizing that a fly had surreptitiously made its way into the firmament, labeled it *Apis* (the bee) on his illustrated plate of the twelve new southern constellations in his 1603 *Uranometria*—then the world's leading star atlas. Indeed, although Plancius's rival, another Dutch cartographer called Willem Janszoon Blaeu, had even given the winged creature its proper Latin name Musca on a 1602 globe, the flat-footed astronomer dug his heels in until 1612, when at last he acknowledged the fly—but only by its Greek name, *Muia*. Despite Plancius's grudging about-face, it continued to be known to some as a bee for a couple of hundred years, as well as Musca Australis (there was also a northern fly, Musca Borealis, now defunct), until it finally established itself in its present incarnation. That's the thing about flies. You can't get rid of them.

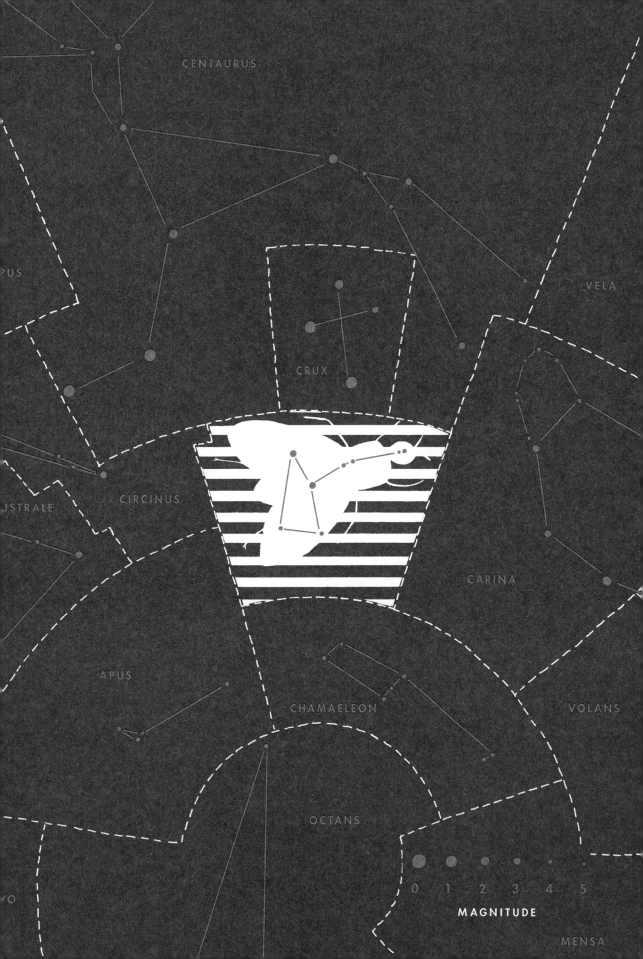

CENTAURUS

VELA

CRUX

PUS

CIRCINUS

JSTRALE

CARINA

APUS

CHAMAELEON

VOLANS

OCTANS

MENSA

MAGNITUDE

0 1 2 3 4 5

NORMA

NOR/NORMAE, THE SET SQUARE

RANK IN SIZE: **74**
ASTERISMS: **NONE**

THE FIRST TIME I saw Mozart's *The Magic Flute*, I was underwhelmed. Or, rather, I was overwhelmed by the amount of money that must have been spent on the extravagantly gauche costumes and set, but I failed to be convinced by either the plot or the merit of the nonsensical opera the English National Opera was endlessly peddling out revivals of. The second time I saw it—probably because I am getting old, and so my brutalized emotions are not affronted by the disingenuity of operatic psychology—I was entranced. This fantastical *Singspiel* extolling the virtues of Freemasonry no longer felt ludicrous. Here (or, rather, down there—a long way from my standing ticket in the gods) were magic, morality, and monsters. The stuff of myth sung by a stellar cast.

Norma (the constellation, not the Bellini opera) is, on the contrary, the stuff of the everyday. There is some historical consternation about what exactly it is supposed to represent—a draftsman's instrument, a surveyor's level, a ship's carpenter's set square—but, despite possible allusions to voyages of exotic exploration, there is hardly anything magical about yet another of the technical instruments in **Lacaille**'s celestial toolbox.

Or so I thought until I considered that there might be another reason why the eighteenth-century astronomer had chosen to set Norma among the stars and, in particular, nearby another of his constellations, Circinus. Joined together, these two symbols—the set square and a pair of compasses—make the emblem of Freemasonry. Was Lacaille's imagery in the sky a conscious Masonic mark? Judging by the acquaintances he kept, it is very tempting to think so. Both his pupils Jean Sylvain Bailly and **Antoine Lavoisier** were Freemasons, while his colleague, the astronomer (and later director of the Paris Observatory) Jérôme Lalande founded the Masonic lodge *les Neuf Soeurs* and was an active member. As indeed were many thinkers (not to mention composers) of the time.

Like its French counterpart, the Royal Society in England (the national academy for science) had a huge roll call of Masons. Sometimes called the "Super Enlightenment," an international thirst for the numinous represented a refusal to reject the unknown, while still being passionately committed to scientific rigor and learning. Eighteenth-century Freemasonry was, despite its accompanying intellectual and moral fervor—and Lacaille was a man prized for his moral rectitude—rife with mysticism and secrecy: just the breeding ground for a theory of reading messages in the stars. Or indeed writing an opera about the virtues of a secret order. Norma has far more in common with Mozart's Masonic masterpiece than I could ever have imagined.

OPHIUCHUS
& SERPENS

LIBRA

HYDRA

SCORPIUS

LUPUS

CENTAURUS

ARA

TRIANGULUM
AUSTRALE

PAVO

CIRCINUS

APUS

MAGNITUDE

0 1 2 3 4 5

OCTANS
OCT/OCTANTIS, THE OCTANT

RANK IN SIZE: **50**
ASTERISMS: **NONE**

EIGHT THINGS ABOUT THE OCTANT TO REFLECT ON

1   In order to construct a celestial version of this navigational instrument—which occupies an appropriate position at the South Celestial Pole—its inventor, **Nicolas Louis de Lacaille**, had to steal some stars from the already lesser water-snake (Hydrus).

2   Appropriately for a device that reflects the path of light to the viewer—and therefore doubles the angle measured—the octant appears to have two simultaneous inventors: an English mathematician from London's Bloomsbury, John Hadley (1682–1744), and a colonial glazier from Pennsylvania, Thomas Godfrey (1704–49).

3   For some unknown reason, although Edmond Halley (1656–1742) of comet fame knew the details of an earlier octant invented by **Isaac Newton**, he took this information about the first reflecting quadrant to his grave.

4   Not to be confused with an octave (music), an octet (music again), an octagon (math), or an octopus (messy).

5   It is the forerunner of the sextant, but whereas that superior navigational tool has an arc of a sixth of a circle, the octant has one—as its Latin name meaning "an eighth" would suggest—of 45°.

6   A Viking would not have used one. Although he might have used crystals to determine the position of the Sun, which, in accordance with his people's astronomical nous, enabled them to calculate their nautical heading.

7   In Egyptian mythology, the Ogdoad are eight creation deities (a quadrant of male gods and a quadrant of female ones) that produced an egg from the primeval waters, out of which the sun god Atum was born. This has got nothing to do with Octans (yet another of Lacaille's legendless constellations). Or indeed the octant.

8   In 1949–50, an undercover CIA agent, Douglas S. Mackiernan, made his way through the Takla Makan desert and a Himalayan winter. To check how accurate his compass was, he constructed an octant out of an old camera. Finally arriving at the Tibetan border, he was shot to death in the snow.

CRUX

MUSCA

CARINA

VOLANS

CHAMAELEON

MENSA

APUS

HYDRUS

TUCANA

PAVO

INDUS

0 1 2 3 4 5

MAGNITUDE

OPHIUCHUS & SERPENS

OPH/OPHIUCHI, THE SERPENT BEARER
SER/SERPENTIS, THE SERPENT

RANK IN SIZE: **11/23**
ASTERISMS: **THE BULL OF PONIATOWSKI**

Incensed with indignation Satan stood
Unterrified, and like a Comet burned,
That fires the length of Ophiuchus huge
In th' Artick Sky, and from his horrid hair
Shakes Pestilence and War.
John Milton, *Paradise Lost*

THE MAN-TURNED-GOD holding the vast snake Serpens aloft in the sky is **Asclepius**, god of medicine and healing. His father was **Apollo** and his mother was **Coronis**, one of the many beautiful mortals the unmarried deity merrily impregnated. Unfortunately for Coronis, Apollo was not as fond of his lovers conducting other romances as he was of having them himself. He had left a white crow to keep an eye on his conquest—Apollo's baby growing in her belly—and so when the lovely Coronis was unfaithful (with a more reliable lover, a mortal called Ischys), the crow flew to his master to tell him of her treachery.

Spitting with the blame-seeking confusion of rage, Apollo did more than shoot the messenger: he cursed the snow-white crow so hard that its feathers, and the feathers of all crows to come, turned blacker than black. But jealous as he was, he couldn't bear to take revenge on Coronis with his own hands; and so he left it to his sister, the huntress **Artemis**, to deliver the fatal arrow-storm from her mighty bow. The mortals mourned Coronis's death, but only after her family had lifted her body onto the funeral pyre, and the smoke began to catch her flesh, did Apollo awake from his rage. Racked by remorse, he dived into the fire, and tearing apart his lover's flesh as it melted into the flames, he ripped the howling baby from her womb.

Apollo handed his son Asclepius into the care of wise **Chiron**, the centaur. Chiron knew best how to raise a boy: the great **Aeneas, Jason, Perseus**, Hercules, **Achilles**, and Ajax were just a handful of the heroes he had tutored in the ways of hunting, music, medicine, and prophecy. But he taught the motherless Asclepius a little too well. So skilled was the boy in the art of healing that he could even raise the dead. He saved Glaucus, Hippolytus, **Tyndareus** (the mortal father of **Castor**), and many more from the clutches of death, and each time **Hades**, king of the underworld, got a little bit more annoyed. Who was this bearded healer stealing souls otherwise destined for him? Eventually the god grumbled, godlily, to his brother **Zeus**, who zapped poor Asclepius with a thunderbolt. Outraged, Apollo then killed the three Cyclopes who made Zeus's thunderbolt in revenge. Eventually, to settle their divine fraternal feud, Zeus made Asclepius immortal and gave him the constellation Serpens to hold forever in the stars.

In Greek Ophiuchus means "toiling," and the Hellenic writers Aratus and Manilius saw the snake as being coiled around Asclepius; but it is unclear as to why he should be pictured in the sky struggling with the very creature that is the instrument and emblem of his powers. It is not just because the snake sheds its skin and is a symbol of rebirth that it is associated with this famous healer. It was a snake that first showed Asclepius how to resurrect the dead. **King Minos**'s son Glaucus, who had fallen into a jar of honey and drowned, lay dead on the ground. As Asclepius attempted in vain to resuscitate him, he saw a snake coming toward them in the grass, and naturally, he killed it with his staff. To his amazement, another snake appeared with a strange herb in its mouth, and placing it on the dead snake's corpse, it brought its fellow creature back to life. Asclepius did the same to Glaucus, and the boy was miraculously revived. This is why the serpent-entwined Rod of Asclepius is a universal symbol of medicine and is used to this day as the logo of emergency services worldwide.

VULPECULA

SAGITTA

HERCULES

AQUILA

SCUTUM

SAGITTARIUS

SCORPIUS

CORONA
BOREALIS

BOÖTES

VIRGO

LIBRA

LUPUS

MAGNITUDE

0 1 2 3 4 5

ORION

ORI/ORIONIS, THE HUNTER

RANK IN SIZE: **26**
ASTERISMS: **THE BELT, THE BUTTERFLY, THE HEAVENLY G, THE RAKE, THE SWORD, THE THREE KINGS, VENUS'S MIRROR, THE WINTER OCTAGON, THE WINTER OVAL, THE WINTER TRIANGLE**

THE DAYS DRAW IN, the pumpkins and the fireworks come out, and the giant hunter Orion appears in the sky, his dogs Canis Major and Canis Minor snapping at his heels. Or at least, he's supposed to. Perhaps as a child I was too busy sparkling my name in the air to notice the glittering of the great hero's belt. Why, as the Victorian writer Thomas Carlyle asked, "did not somebody teach me the constellations, and make me at home in the starry heavens, which are always overhead, and which I don't half know to this day?"

In days of old, mothers sang lullabies to their children to pass on their knowledge of the stars. I wish that as I'd lain in my sticker-clad bunk bed, one of my many babysitters had told me about the stellar nursery hidden just south of this huntsman's belt: the Orion nebula in which not just stars but whole solar systems are formed. Or about Shen, the warrior who—in a rare conjunction with Western astronomy—the Chinese also see in these stars, as part of a celestial hunting tableau. Why, when I wrote a whole project on Inuits, did I not learn that they, too, saw hunters chasing through the night? Why did my New Zealand au pair not tell me about the Canoe of Tamarereti, the mythical ancestor of the Maori people? Or my Norwegian nanny not thrill me with tales of Thor breaking off Aurvandil's frozen toe and casting it up into the heavens?

When someone plunked me down in front of Tim Burton's 1988 film, I had no idea that the devious hero of this spooky comedy owed his name to the unpredictable variable supergiant, Betelgeuse. Nor that this star marking the hunter's right shoulder, although designated as Orion's α star, is in fact only its second most brilliant. The constellation's lucida is actually Rigel, from the Arabic rijl meaning "foot"; but this "bright star in the left foot," as Ptolemy called it, is labeled β Orionis. Nor was my imagination sparked by the fact that a female warrior—the supergiant Bellatrix, sometimes known as the Amazon star— sits on Orion's left shoulder.

Only in adulthood have I learned that this star-spangled hunter is the anthropological descendant of the great Sumerian hero Uru An-na (meaning "light of heaven"), who fought the Bull of Heaven and whom we call **Gilgamesh**; and that this is why Orion is still pictured brandishing his club and lion pelt against the charging Taurus. And learned to see in this image the undeniable hints of another ancient hero, Hercules. And that there are so many stories to explain the presence of this glorious giant, the most recognizable of all constellations, that I would need several childhoods to listen to them all in awe.

GEMINI

TAURUS

MONOCEROS

ERIDANUS

CANIS MAJOR

LEPUS

0 1 2 3 4 5

MAGNITUDE

PAVO

PAV/PAVONIS, THE PEACOCK

RANK IN SIZE: **44**
ASTERISMS: **NONE**

Juno's bird, which wears stars on its tail
Ovid, *Metamorphoses XV*

JUNO has a hard time. Despite (or perhaps because of) being the goddess of marriage and the matron of all Roman women, the ever-spurned spouse of **Jupiter** spends most of her time dealing with her husband's infidelities; and yet *she*'s the one who gets the bad press. The Greeks (who call her **Hera**) are forever casting her in the role of the whining, vengeful shrew, while **Virgil** depicts her as savage and cruelly vindictive, laying pretty much all the blame for **Aeneas**'s trials and tribulations at her door. It is true that sometimes she goes a little far when inflicting revenge on Jupiter's impregnated paramours. (You need only look to the stars of Ursa Major to see what she did to sad **Callisto**, or to the constellation Taurus to see her harsh punishment of the innocent **Io**.) No wonder modern authors have used the much-maligned goddess as a sort of emblem of domestic unbliss. Sean O'Casey's play *Juno and the Paycock* translates the divine comedy of marital discord into the very mortal tragedy of a Dublin family from a squalid tenement block torn apart by Ireland's civil war in the 1920s. At least the 2007 film *Juno*, with its welcome subversion of ancient morality tales and surprisingly happy ending, reclaims some hope and dignity for the world of unexpected pregnancies and unrequited love.

Despite the tough time she has of things, Juno does at least get to ride around the heavens in a chariot drawn by peacocks. The peacock, as the Dutch navigators **Keyser** and **de Houtman**, who placed one in the stars of the southern hemisphere, were well aware, was her sacred bird. You will remember from Io's story that when Juno's hundred-eyed guardsman **Argus**—and no wonder she needed one—was slaughtered by **Mercury** (who I called in that story by his Greek name, **Hermes**) under Jupiter's command, she placed his eyes in the peacock's feathers. But perhaps it is not just this etiological myth that explains why this bird should be so commonly associated with this willful goddess.

The peacock, as every disappointed little girl one day discovers when admiring one over the fence of a city park (as it happened to me), is a man. And the lady peacock, who is in fact called a peahen, is that boring gray-feathered lump over there, pecking at the last crumbs of bread from the handful you tossed excitedly in the direction of the other, astonishing rainbow of a creature five undisappointed minutes ago.

Today when I look at peacocks (on the Internet, mostly) I think of Juno, looking on while her husband struts around, flaunting his stuff, seducing young girls.

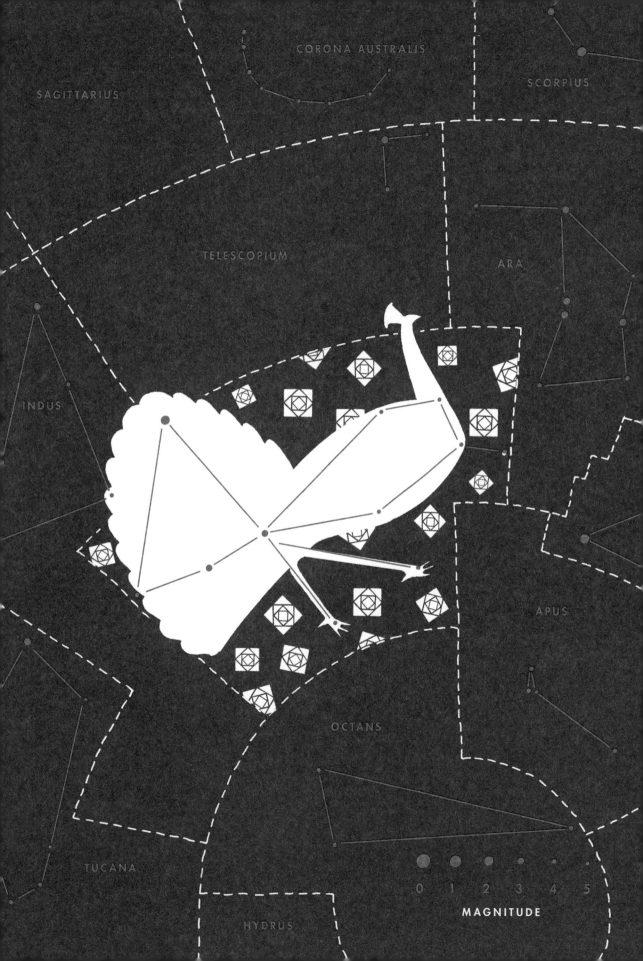

SAGITTARIUS

CORONA AUSTRALIS

SCORPIUS

TELESCOPIUM

ARA

INDUS

APUS

OCTANS

TUCANA

HYDRUS

MAGNITUDE

0 1 2 3 4 5

PEGASUS

PEG/PEGASI, THE WINGED HORSE

....

RANK IN SIZE: **7**
ASTERISMS: **THE BASEBALL DIAMOND, THE GREAT SQUARE,**
THE LARGE DIPPER

....

WHAT LINKS Daphne du Maurier, **Medusa**, a bridge in Normandy, horse-whispering, British paratroopers, **Poseidon**, and **John Keats**? The answer, as I'm sure you've guessed, is in the stars.

Is it September? About midnight? Good. Go outside and look up. Find the blue-white star at the top of Andromeda's head (α Andromedae). Can you see it? Excellent. This star was once known as Sirrah, meaning "navel" in Arabic, and marks that very point on the creature we're looking for. Now trace a line toward the western horizon to the deep-yellow star, β Pegasi—known either as Scheat (shin) or Menkib (shoulder). From there, draw a line southward to α Pegasi—Markab, the saddle. Extend it toward the eastern horizon to join γ Pegasi or Algenib (the side), then continue north, and you will find yourself back at Sirrah, (or Alpheratz, as α Andromedae is now usually known), having drawn the Great Square of Pegasus. This famous asterism forms the body of the winged horse we're looking for. Its neck is marked by Homam (ζ Pegasi), whose name means "horse-whisperer" and evokes the power of that mysterious and ancient magic used to tame wild stallions and mares. If you find the yellow supergiant Enif (ε Pegasi), you will have found its nose. Perhaps it wears the golden bridle that **Athene** brought to the hero Bellerophon in a dream, so that he might tame that magical horse. For it was on Pegasus that Bellerophon took to the sky, lance in hand, and swooped down on the Chimaera, a fire-breathing lion-snake-goat.

When British paratroopers first took to the sky in 1941, their commander, Lieutenant General Sir Frederick "Boy" Browning, chose this ancient airborne warrior as the insignia the 1st Airborne Division wore on their upper sleeves. Three years later, it graced the arms of the 6th Airborne Division on the night they captured a vital bridge over the Caen Canal in the beginning moments of the invasion of Normandy; taking this bridge was absolutely crucial to Operation Deadstick (as the airborne forces' operation was code-named), and it has been known ever since as Pegasus Bridge. There is some debate as to whether the iconic image of a light-blue Bellerophon riding Pegasus on a maroon background was designed by the artist Edward Seago, or by the commander's wife, the novelist Daphne du Maurier. Nevertheless, this symbol of the airborne forces is still worn to this day (alongside the famous maroon cap that Sir Frederick also introduced) by Britain's Parachute Regiment, in operations from Iraq to Sierra Leone.

Despite Bellerophon's military credentials, Pegasus was in fact a peaceful horse. Associated in antiquity with poetic inspiration, it was his hooves that upturned the turf on Mount Helicon to create the gushing spring sacred to the nine muses that the poet Keats so longed to drink the waters of:

> *O for a draught of vintage! that hath been*
> *Cool'd a long age in the deep-delvèd earth,*
> *Tasting of Flora and the country-green,*
> *Dance, and Provençal song, and sunburnt mirth!*
> *O for a beaker full of the warm South!*
> *Full of the true, the blushful Hippocrene,*
> *With beaded bubbles winking at the brim,*
> *And purple-stainèd mouth;*
> *That I might drink, and leave the world unseen,*
> *And with thee fade away into the forest dim.*

Yet although he was the creator of such a tranquil fountain—the name Hippocrene means "horse's fountain"—this divine steed had a violent birth. When Poseidon disguised himself as a horse to seduce a flowing-haired Gorgon virgin called Medusa in the temple of Athene, the enraged goddess transformed the young beauty into a snake-tressed monster whose glare turned men to stone. It was only much later, when Perseus slew Medusa, that Pegasus, the creature they had conceived, sprang from her decapitated body.

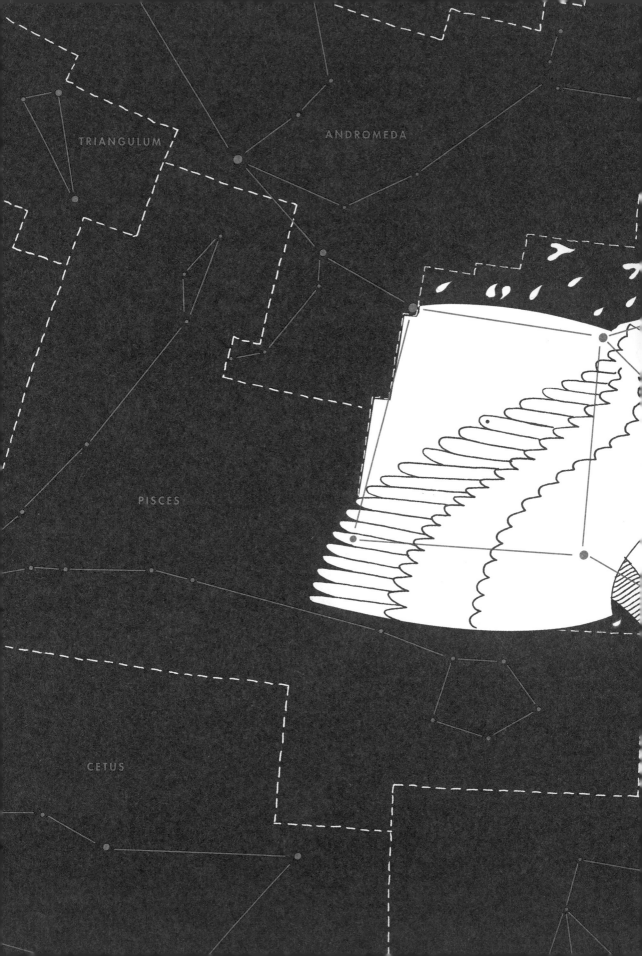

TRIANGULUM

ANDROMEDA

PISCES

CETUS

LYRA

LACERTA

CYGNUS

VULPECULA

SAGITTA

AQUILA

DELPHINUS

EQUULEUS

AQUARIUS

0 1 2 3 4 5

MAGNITUDE

PERSEUS

PER/PERSEI, THE HERO

· ·

RANK IN SIZE: **24**
ASTERISMS: **THE LARGE DIPPER, THE SEGMENT**

· ·

CELESTIAL TELEGRAM SERVICE

To: **Perseus**, *mortal son of* **Zeus**
From: King Polydectes, Island of Seriphus

Your mother Danaë and I getting married STOP Personally think it's none of your
business STOP But you've always been a mommy's boy STOP However will stop pursuing
her on one condition STOP Bring me the head of the Gorgon **Medusa** STOP

To: King Polydectes, Island of Seriphus
From: Perseus, mortal son of Zeus

You well know Polydectes that is an impossible task STOP But shall not be dissuaded STOP
My mother as sick of your advances as I am STOP **Hermes** has given me a sharp sickle to
execute the old hag STOP **Athene** has given me polished shield to avoid her gaze STOP
You're on STOP P dot S dot STOP I have come top of my class in discus and wrestling
every year STOP I'm no mommy's boy STOP

To: Perseus, mortal son of Zeus
From: Hermes, Olympus

Have faith Perseus STOP We gods will give you helmet to make you invisible and winged
sandals and magic pouch STOP Go find the three Graeae in royal palace on Atlas's mountain
STOP Look for nymphs with bodies of swans STOP They are the sisters of the Gorgon
Medusa and know where she lives STOP Have one eye and one tooth between them
which they pass back and forth STOP Good luck STOP

To: King Polydectes, Island of Seriphus
From: Perseus, mortal son of Zeus

Medusa slaughtered STOP Tricked her blind sisters into giving me directions STOP Crept up
on her while she was asleep STOP Got her head in my leather pouch STOP Not as hard as
imagined STOP At moment she died winged horse arose from her body STOP Flying home
on it over Ethiopia heard screams STOP Saw beautiful woman called Andromeda chained
to rock and in much distress STOP Said if I could slay beast about to eat her she'd marry
me STOP Quickly asked father's permission for her hand and rescued her STOP Turned sea
leviathan to stone with Medusa's head STOP Looks like I'm getting married and you're not
STOP Mommy's boy eh question mark STOP

CASSIOPEIA

CAMELOPARDALIS

AURIGA

ANDROMEDA

TRIANGULUM

ARIES

TAURUS

0 1 2 3 4 5

MAGNITUDE

PHOENIX
PHE/PHOENICIS, THE PHOENIX

RANK IN SIZE: **37**
ASTERISMS: **NONE**

Una est, quae reparet seque ipsa reseminte, ales:
Assyrii phoenica vocant . . .

WHO CAN TELL if **Keyser** and **de Houtman** were thinking of **Ovid** when they placed a Phoenix—that ancient symbol of immortality and resurrection—in their astral menagerie?

Here is my translation of what Pythagoras teaches the Romans about the Phoenix in the last book of Ovid's *Metamorphoses*.

...

There is one.
One bird.
Which renews
itself, reproduces
itself, regenerates
itself—SPRINGS!—*from*—
itself.
The Assyrians call it
The Phoenix.

It doesn't live on crops or grasses or anything as ordinary and boring as that but on the TEARS OF FRANKINCENSE and the magic potion that is the juice in the sap of a spice plant from the east. It lives five generations to the full. Then. Immediately. Using only its talons and its innocent beak, it builds itself a nest in the shaking branches of a palm tree's summit. As soon as it has laid a scented bed of cassia bark, smooth spikes of muskroot, tawny myrrh, and broken cinnamon

It lays
itself
on top of its odorous plot
and finishes
that cycle
of its life.
And from the body
of the father phoenix
A little Phoenix
is reborn.

CETUS

SCULPTOR

FORNAX

ERIDANUS

GRUS

LOGIUM

TUCANA

ULUM

HYDRUS

0 1 2 3 4 5

MAGNITUDE

PICTOR
PIC/PICTORIS, THE PAINTER'S EASEL

RANK IN SIZE: 59
ASTERISMS: NONE

The colours of the stars are not of course as those of our painting; they are transparent and luminous; in order to reproduce them we would have to have as our palette the azure of the skies and be able to dip our brush in the rainbow.
Camille Flammarion (1842–1925)

IN 1756, if you were to open the *Mémoires de l'Académie Royale des Sciences* published that year, you would find an exquisite engraving of **Lacaille**'s star chart illustrating all fourteen of his new constellations. Including all of **Ptolemy**'s original figures—as well as the newcomers to the celestial globe added by **Keyser, de Houtman,** and **Petrus Plancius**—this painstakingly accurate yet aesthetically gorgeous planisphere reconfigured the heavens according to the vision of the indefatigable French astronomer. (Lacaille had in several cases to fiddle with the firmament to incorporate his own inventions—for example cropping the tail of Hydrus to make space for his Octans, or stealing a couple of that water-snake's stars to fill out the faint figures of Horologium and Reticulum).

Among the beautifully rendered gods, animals, and scientific instruments was *le Chevalet et la Palette*—the Easel and the Palette. This topographical tribute to the skill and value of the painter is evidence of the high esteem in which artists were held in Lacaille's enlightened age. In the second edition of his planisphere, published posthumously in his star catalogue *Coelum Australe Stelliferum* in 1763, he Latinized the constellation to Equuleus Pictorius [*sic*]. As so often, the name was further tinkered with by subsequent astronomers—the German astronomer **Johann Bode** called it Pluteum Pictoris—before settling on its shortened moniker.

Though faint and insignificant, populated only by minor stars, Pictor serves as a metaphor for the astronomer's work as well as a testament to the artist's toil. With celestial cartography, scientists paint the sky: their chance—to borrow a phrase from the French astronomer Camille Flammarion—to dip their brush in the rainbow. From **Johann Bayer**'s first great celestial atlas in 1603, to John Flamsteed's 1729 *Atlas Coelestis*, to Bode's splendid 1801 *Uranographia*, these maps were not just ways to transmit their astonishing observations and discoveries but spectacular and highly popular works of art in themselves. (Far more accessible, too, than expensive, individually handcrafted globes.) These works of ingenuity would have been as thrilling to contemporary audiences as the Hubble Space Telescope images that we Google today.

Wonderfully, all this aesthetic reconfiguration is in keeping with the behavior of the stars themselves. In 1984, a photograph revealed a disc of material surrounding the blue-white β Pictoris, suggesting that this star is busy forming an entirely new planetary system: painting its own vision on the universe.

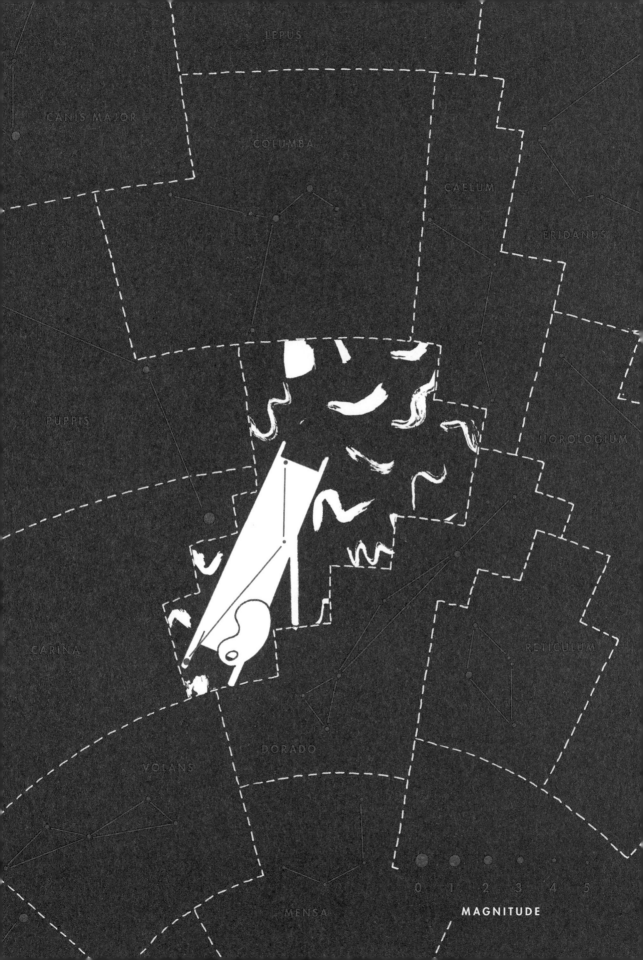

LEPUS

CANIS MAJOR

COLUMBA

CAELUM

ERIDANUS

PUPPIS

HOROLOGIUM

CARINA

RETICULUM

DORADO

VOLANS

RETICULUM

MENSA

0 1 2 3 4 5

MAGNITUDE

PISCES
PSC/PISCIUM, THE FISHES

....

RANK IN SIZE: **14**
ASTERISMS: **THE CIRCLET**

....

TODD HAD been driving down an empty freeway for hours, the tinny drone of an indeterminate country music station worming its way into his subconscious. This car in front of him was the first he'd seen all day. A rusty estate driving up the dust in its wake, it was moving through the pale, parched landscape at unusual speed. So he was surprised to notice not just a woman at its wheel but a fish bumper sticker at its rear. Excited by some company on the road, he put his foot down on the accelerator and overtook her with a grin, hoping that she would notice the sign on the back of his truck and honk if she loved Jesus.

She didn't honk. Glancing behind him in his mirror, he saw a dark-haired woman in her midthirties, hair scraped back behind a worn face, staring straight ahead—her eyes seemingly fixed on, but clearly not looking at, the road. Disappointed, he sped into the late September sun.

Carla had hardly noticed the truck until she heard a gurgled "Chuck! Chuck!" and the sound of a small fist thumping against the window behind her. Billy still seemed to have no sense of what had happened and was smiling in delight at a universe he had not yet discovered would go on to treat him pretty badly. The pain of this hit Carla in the chest with a force that startled even her—a woman used to unpredictable and indiscriminate violence. For the first time in those past two days, she began to sob. Uncontrollably. She pulled over, turned off the ignition, and stumbled out of the car as quickly as she could.

She must have been standing there for hours. It was dark, and she was shivering: a clear night, she thought, looking up.

She had never seen anything like it. Above her the night was pricked with lights so sharp it was as if everything she had ever known had been completely out of focus, and all the sense and mystery of the world was now inscribed on the darkness in an entirely new language in an unknown alphabet.

Later, procrastinating from the accounting course she did online after her shifts at the diner when Billy was asleep, she would discover that she had been standing under the stars of Pisces; and that one of the two fishes that swam in her gaze was **Venus**, and that the other was the goddess's son, **Cupid**. Scouring the Internet for anything she could find about astronomy and space, she came across asteroids, aliens, and the ancient story of how this divine mother and son ended up in the stars. Of how they had fled from the ferocious **Typhon**, a terrifying monster sent by Mother Earth, who was called **Gaia**, to attack the gods. Beset upon by this beast, whose eyes blazed with fire, and who yelled and howled from his many heads in voices that changed from one day to the next, the mother and son had escaped to the banks of the Euphrates and hidden in the reeds. Seeing their fellow god **Pan** jump into the river (this is how he became the goat-fish Capricornus), and hearing rustling behind her, Venus called to the water nymphs for help. But before they could heed her cries, Typhon caught sight of her and Cupid and made straight for the riverbank, flames in his eyes, and black tongues flicking wildly from his hundred heads. Tying a cord around her son's ankle so they would not lose each other, Venus took the boy in her arms and dived in. As they broke the water's surface, they turned into fish, darting through the speeding stream.

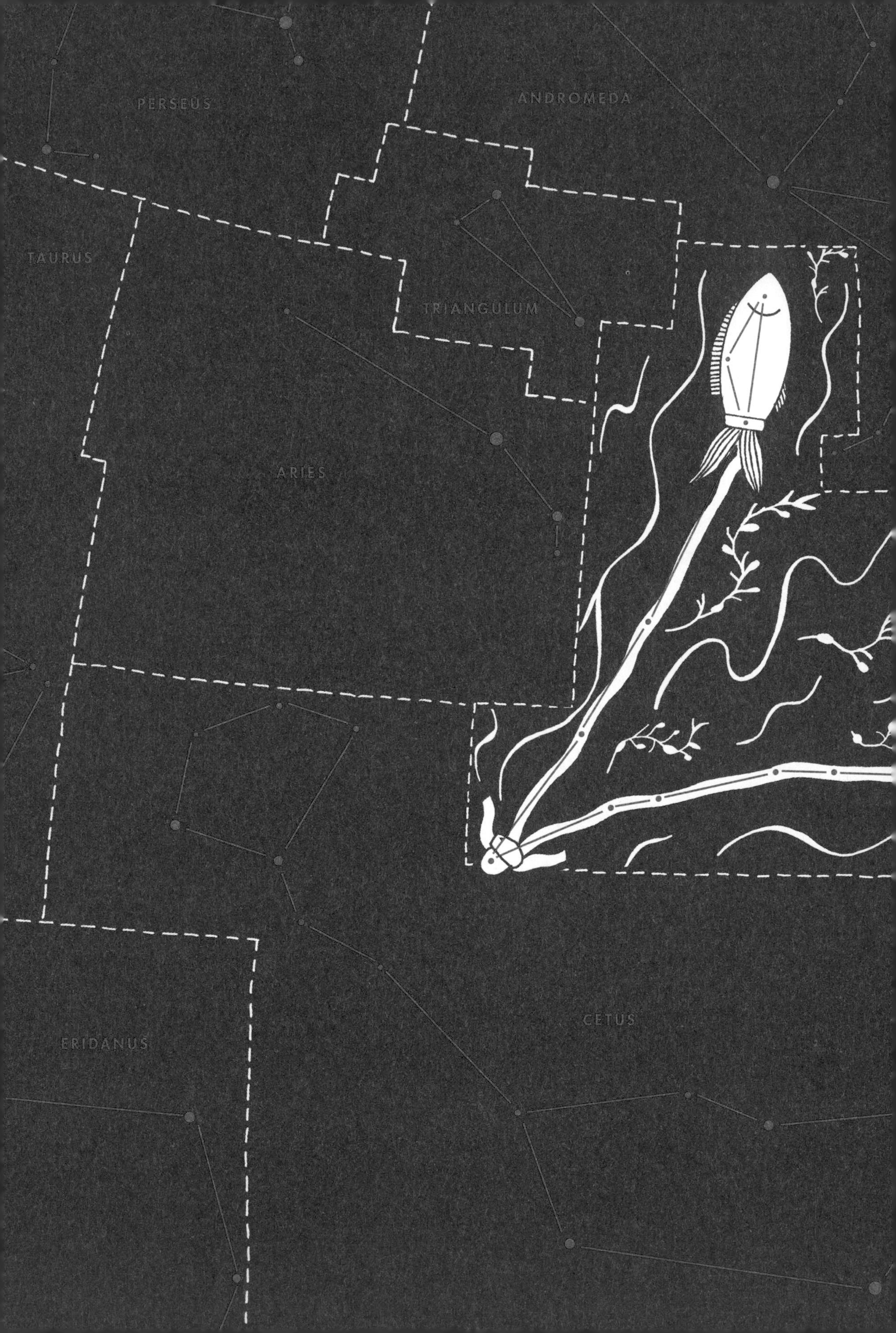
PERSEUS
ANDROMEDA
TAURUS
TRIANGULUM
ARIES
CETUS
ERIDANUS

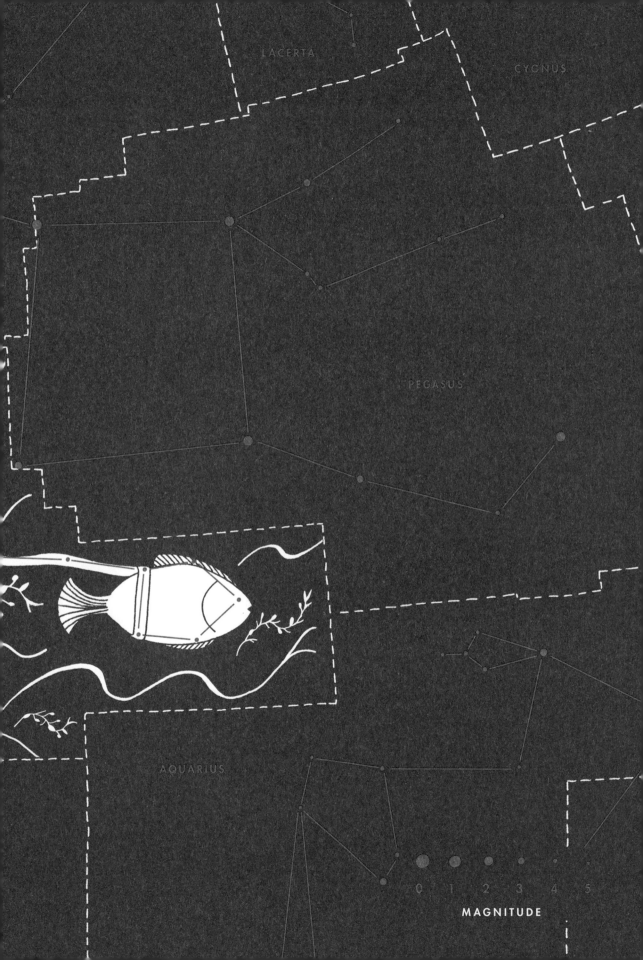

LACERTA

CYGNUS

PEGASUS

AQUARIUS

● ● ● · · · ·
0 1 2 3 4 5

MAGNITUDE

PISCIS AUSTRINUS

PSA/PISCIS AUSTRINI, THE SOUTHERN FISH

RANK IN SIZE: **60**
ASTERISMS: **NONE**

ISHTAR IS DRAWING a mermaid. She'd never heard of one before: she lost all the ways that she might have heard—school, television, her mother and her two sisters—when her family's house was destroyed by heavy shelling. But there's a redheaded one with a green tail, a purple-shelled bikini top, and giant blue eyes smiling out at her from the T-shirt her cousin is wearing. (Aaliyah was given it on the very first day they came here to the camp, and both girls had been astonished when the kind lady with the blue sweater had told them it had come all the way from Germany. Why would any child in her right mind have given up such a treasure? The cousins came up with endless, ever more far-fetched fantasies about the tragedy that must have befallen this unfortunate German girl.)

Halfway through the tail, her crayon breaks in half. She feels as though she wants to cry, but as usual no tears seem to come out. The tired-looking man with the red gym shoes and no hair comes over and gives her another one, but it's the wrong color. Then he sits next to her and starts cutting out stars from a yellow piece of card stock. She sticks them above the half-formed mermaid's head. It's been a long time since Ishtar looked up at the stars. For the past two and a half years she has spent most of her waking hours, and all of her sleeping ones, terrified by the things that fall from the sky. It is of little consequence to her that the home she fled in Manbij, a town in the Aleppo Governorate, was known to her ancestors as Bambyce, and as Hieropolis ("the Sacred City") to the ancient Greeks. Or that her birthplace some thirty kilometers west of the Euphrates was also the center of a cult of worship of the Syrian fertility goddess, Atargatis. It is unlikely that either Ishtar or Aaliyah will be told any one of the several fishy tales about that ancient deity. Or that they will debate with as much enthusiasm the conflicting myths about how, whether, or why Atargatis fell into a lake and was either saved by or turned into a fish, as they did the origins of that secondhand *Little Mermaid* T-shirt.

Ishtar takes her drawing and pastes it to the wall. Looking at it amid all the other girls' pictures, it seems lost.

AQUARIUS

CAPRICORNUS

SCULPTOR

MICROSCOPIUM

GRUS

PHOENIX

0 1 2 3 4 5

MAGNITUDE

PUPPIS

PUP/PUPPIS, THE STERN (OF ARGO NAVIS)

RANK IN SIZE: **20**
ASTERISMS: **NONE**

JASON AND THE ARGONAUTS: AN EPIC TALE

*(in which fifty Hellenic heroes set sail for Colchis in a fifty-oared ship to bring
the Golden Fleece and the ghost of* **Phrixus** *back to Greece)*

TOLD BACKWARD IN THREE PARTS

(much as **Ptolemy**'s *original constellation Argo Navis, which honored this legendary ship,
was split into three parts—Carina,* Puppis, *and Vela—by the French astronomer* **Lacaille** *in 1756)*

PART II: PUPPIS

AS THE DRAGON hisses and uncurls itself toward him, seething through all of its thousand coils, **Jason** sees that it is vaster even than the *Argo* itself. **Medea** steps forward. She looks the dragon in the eye and begins to incant in a strange tongue. He finds himself burning with lust. How strange that in this moment, as she wields her juniper sprigs in front of the scaled and fuming face and pours strange liquids on its drooping eyes, he should finally fall for her. This woman who has hurled herself in love at him; who has scorned her own father's wrath in helping him perform the near-impossible tasks **King Aietes** has set; who has bathed him in the bloodred sap of a saffron crocus so that he could yoke the king's fire-breathing bulls; who has led him and his Argonauts here to the sacred grove in the depths of the night; but to whom until now he has felt only the most expedient gratitude.

As he stands here facing the last obstacle of his voyage, he thinks of all the other ordeals that he and his death-diminished crew have overcome. The arrow-feathered birds that hailed down on them as they approached this island's shores. The foul food-snatching, excrement-strewing harpies from whom they rescued the blind seer Phineus; the clashing rocks, those dreaded Symplegades that Phineus then told them the way through. He thinks of **Polydeuces** boxing the savage Amykos, splintering his skull to slivers of bone. He thinks of Heracles searching the woods of Mysia for his slave and lover Hylas—who had been sucked into a spring by a forest nymph—the hero, howling in grief.

He does not yet know that his journey home will be haunted by the song of the Sirens, drowned out only by the intoxicating sound of **Orpheus**'s lyre. That only tomorrow, as King Aietes's fleet pursues the escaping *Argo*, he will murder Medea's brother Apsyrtus: that he will cut off the boy's hands, feet, nose, and ears and then lick his blood and spit it out three times to prevent his ghost from having its vengeance.

Jason grabs Medea by the hips as she leads him past the sleeping monster and to the oak tree where his gleaming prize hangs.

CANIS MAJOR

PYXIS

COLUMBA

VELA

0 1 2 3 4 5

MAGNITUDE

PYXIS

PYX/PYXIDIS, THE SHIP'S COMPASS

RANK IN SIZE: **65**

ASTERISMS: **NONE**

IN 1269, a man called Peter Peregrinus wrote a letter in Latin promising to teach his "dearest of friends" a thing or two about attraction and repulsion. Little is known about the French scholar who was the author of this work—which, going by the large number of surviving manuscripts, was highly popular in the Middle Ages. His name translates most directly as "Peter, the pilgrim," and one thing we do know is that when he penned this (as it turned out to be) landmark epistle, he was serving in the army of Charles I of Anjou on a crusade sanctioned by the pope. The Latin word *peregrinus* also carries some epistemological echoes appropriate to Peter's treatise. One wonders, when he chose his moniker, whether he was thinking more of its connotations of roving, migration, and wandering than of religious devotion. For Peter's words of wisdom about attraction and repulsion did not constitute a medieval guide to dating, but rather the first extant scientific exploration of magnetism and its use in navigation:

> I will now make known to you, in an unpolished narrative, the undoubted though hidden virtue of the lodestone, concerning which philosophers up to the present time give us no information, because it is characteristic of good things to be hidden in darkness until they are brought to light by application to public utility.

Honorable as this intention was, and as valuable as it turned out to be to the history of Western science, it was based on a fallacy: the philosophers that this soldier-scientist came across may have concealed the magnetic powers of the lodestone, but the ancient Chinese had been harnessing them since the birth of the Common Era. Long before Portuguese explorers went in search of treasure in the East Indies, the Chinese were using this naturally occurring magnet to seek out their own gems. Having discovered that a suspended lodestone will always point in the same direction (toward the magnetic poles), they used it initially not in nautical compasses but in geomancy and fortune-telling, as well as to orient their environments harmoniously in feng shui.

It is not fully known how the invention of the compass made its journey from China to Europe, although many historians believe that it must have been via the trading routes of the Arabs. But whether it traveled north, east, south, or west, it revolutionized the way that seafarers were able to negotiate the waves. Which of course is why the astronomer **Lacaille** set Pyxis—not to be confused with Circinus, the pair of surveyor's compasses he also added to the sky, nor indeed with Lyra's magical alethiometer from Philip Pullman's *His Dark Materials* trilogy—amid the stars.

HYDRA
SEXTANS
MONOCEROS
PUPPIS
ANTLIA
VELA
0 1 2 3 4 5
MAGNITUDE

RETICULUM

RET/RETICULI, THE NET

RANK IN SIZE: 82

ASTERISMS: NONE

IF YOU THOUGHT that the fastidious astronomer **Lacaille**'s penchant for the minute stopped at his installation of a microscope (Microscopium) in the sky, then think again. Reticulum is the constellation that he created to honor the reticle or grid in the eyepiece of the small telescope with which he scanned the stars from near Table Mountain (Mensa), during his trip to the Cape of Good Hope in 1751–2. Although without this deceptively "little instrument"—as he described it in his introduction to his catalogue of the southern stars—he would not have been able to make the huge additions to astronomy that he did.

King Alfonso X of Castile—the author of the thirteenth-century Alfonsine tables of astronomical data—once said, "Had I been present at the Creation, I would have given some useful hints for better ordering of the universe." Not that I begrudge Lacaille his scientific apparatus, or all the men before and since for their curation of the night sky, but I think I would have done things a little differently too. Were I to possess either the astronomical skill, mathematical ability, or even just political muscle to reframe the official eighty-eight internationally agreed constellations, I would correct some serious omissions.

For a start, where's Buddha? Even the National Health Service is coming around to the notion of mindfulness nowadays. And what about Karl Marx, Gandhi, Einstein, or Jung? Technology has accelerated at a ferocious speed since Lacaille's day, and it seems only fair that we should be able to commemorate some of our most life-changing inventions: the computer, penicillin, the garlic crusher, the car. It might be kind to remember the dodo and the dinosaur, as well as other extinct and dangerous creatures like Margaret Thatcher—who could perhaps be depicted stealing milk-bottle asterisms from the Milky Way. I'm not sure why the ancients forgot to place Cassandra in the sky (not even the mythographers listened to her, it turned out), and I would redress their astral whitewashing of Cleopatra and Boadicea. There is a notable lack of poets and writers—not least snubbed is William Blake, who illustrated the constellations so exquisitely. It might be fun to dedicate a black hole to *Tristram Shandy*; Virginia Woolf should surely be allowed a meteor stream of consciousness and even the existentialist Samuel Beckett might enjoy a place in eternity. I'd have a hyena in a petticoat to replace Pavo the Peacock (so as not to forget Horace Walpole's vile condemnation of the mother of feminism, Mary Wollstonecraft), and I think I'd reconfigure the chained maiden Andromeda as a suffragette tied to the railings, or even a vent-straddling Marilyn Monroe.

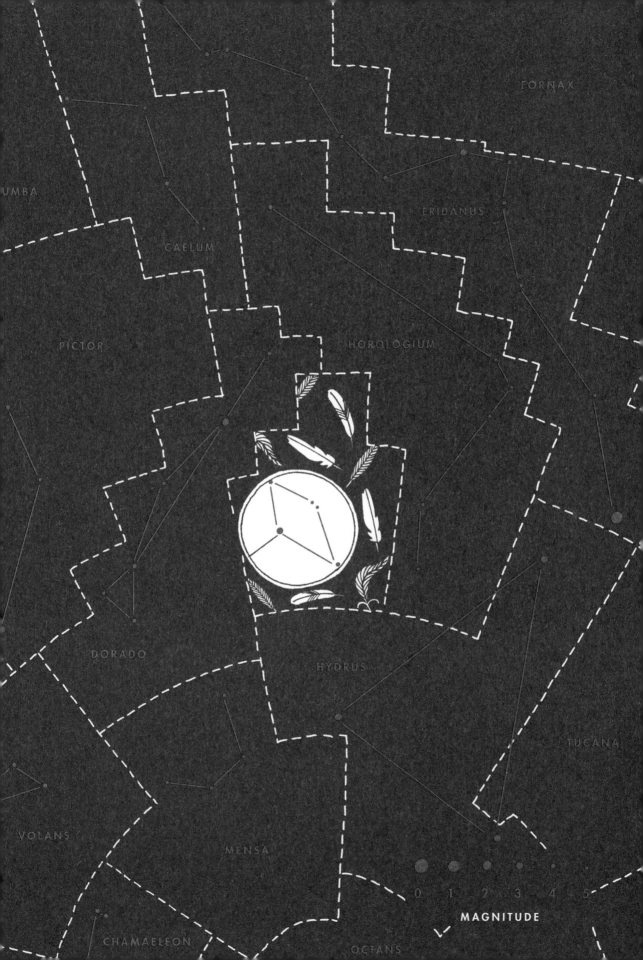

FORNAX

UMBA

ERIDANUS

CAELUM

PICTOR

HOROLOGIUM

DORADO

HYDRUS

TUCANA

VOLANS

MENSA

CHAMAELEON

OCTANS

0 1 2 3 4 5

MAGNITUDE

SAGITTA

SGE/SAGITTAE, THE ARROW

RANK IN SIZE: **86**
ASTERISMS: **NONE**

THERE ARE ARROWS flying left, right, and center across the stories in the sky. Who's to say which way this one is whistling and why? It depends who's looking.

Is it one of the legendary hero Heracles's arrows? (Oh, Heracles, Hercules, you say "tomatoe," I say "tomahto"—take your pick—the Greeks and the Romans had their own transatlantic linguistic rifts). Perhaps it is the one he used to kill Aquila, the eagle sent by **Zeus** to peck at **Prometheus**'s innards, and in doing so save that fire-stealing hero from endless torture. Or perhaps it is one of the arrows with which **Apollo** slayed the **Cyclopes**, to punish them for making the thunderbolts that killed his son **Asclepius**. Perhaps it is even one of **Eros**'s missiles, which pierced the heart of Zeus so deeply that he fell hook, line, and sinker for the nubile shepherd **Ganymede**. Well, never mind if it is. Ganymede is remembered in the constellation Aquarius, and Asclepius in Ophiuchus; and Aquila the eagle told you all about Prometheus himself, so let's have a story we haven't heard. It's one about Heracles, after all.

Do you ever find it slightly disturbing that birds look a lot like dinosaurs? Do you worry that behind the points of those tiny, soulless eyes is a violence latent in their genes? Well, imagine a numberless flock of brazen birds—their wings, claws, and beaks all bronze—that have bred so rapaciously there is not an inch left uncovered on the putrid marsh they inhabit. Worse still, their metallic feathers are tipped with poison, readied to unleash a shower of pain; their excrement is toxic; and they feast on the flesh of men. That is the sight that faced our lion-pelted hero as he arrived on the shores of the Stymphalian river. Well he knew that these birds, which looked like ibises and were the size of cranes, could break a metal shield or breastplate; he knew, too, that in the Arabian desert, where they also bred, they were feared more than leopards and even lions, for they would launch themselves at travelers, transfixing them in agony. It is fair to say that Heracles, staring upon the object of his Sixth Labor, was not a little scared.

Thankfully, the goddess **Athene** came to his rescue. She gave him a brazen rattle to battle these brazen birds. Heracles shook this mighty rattle—which had been welded in the forge of **Hephaestus**, the great blacksmith god—so loud that the entire hideous flock took to the sky. Pursuing them with arrows, Heracles hit one of the Stymphalian birds with such force that it exploded into toxic smithereens. A gruesome scene brought to mind by this astral projectile, poised eternally between the celestial birds Cygnus and Aquila.

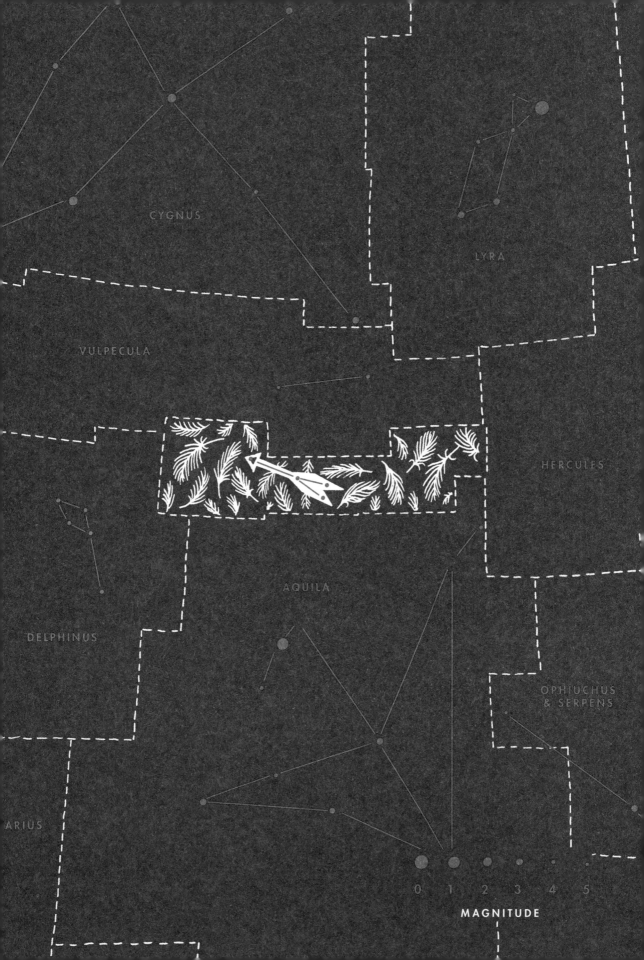

CYGNUS

LYRA

VULPECULA

HERCULES

AQUILA

DELPHINUS

OPHIUCHUS
& SERPENS

ARIUS

MAGNITUDE

0 1 2 3 4 5

SAGITTARIUS

SGR/SAGITTARII, THE ARCHER

RANK IN SIZE: **15**
ASTERISMS: **THE MILK DIPPER**

IF YOU WANTED to shoot an arrow right into the center of the Milky Way, you would aim for a point at the very edge of this constellation—right by the border with Scorpius, close to γ Sagittarii. If, having taken aim, you were then to draw back your bow and stretch it so taut that the projectile you let loose could travel 26,000 light-years, you would inadvertently hit Sagittarius A*, the supermassive black hole at the very epicenter of our galaxy, the radio source consuming dust and gases that acts as its gravitational anchor, the galactic bull's-eye of the Milky Way.

Your arrow would disappear faster than you can say "space–time continuum"—this black hole swallows whole stars for its supper—into a nothingness millions to billions times the mass of the Sun, and who knows what Narnia awaits it there?

There may well be Mr. Tumnus taking Lucy home for tea; there may even be a roasting fire and sardines-on-toast and cake. For Sagittarius is also known as the Teapot: you can see the steam of the Milky Way rising from its spout, and the ladle called the Milk Dipper formed by the stars of Lambda, Phi, Sigma, Tau, and Zeta Sagittarii. Besides, Mr. Tumnus is a faun, and although that's not quite the same as a satyr (both are part-man part-goat, but a faun has a human face), which in turn is not quite the same as a centaur (both have horselike tails but centaurs do not use bows), it's still very much a follower of **Pan**, a sylvan dweller and the sort of creature that has been spied in the stars since Sumerian times.

The Greeks' galactic goat-god was Crotus. A playful satyr who invented archery and applause, he was the son of the nymph Eupheme. Since Eupheme was nurse to the nine muses (Clio, Thalia, Erato, Euterpe, Polyhymnia, Calliope, Terpsichore, Melpomene, and **Urania**, the daughters of **Zeus** and Mnemosyne), she raised her son alongside her divine charges, and they all gamboled about happily on Mount Helicon. The muses would sing and Crotus would sit on a tree stump and listen; and when they had finished he would put his hands together and take them away again and then do it again and again even quicker, making a sound that delighted them all. When their beloved companion died, the muses asked Zeus to place him in the stars. Their father agreed and to spoil his girls even put Cronus's wreath (which would always fall off his head as they played) beneath the archer's feet in the ringlet of stars that is Corona Australis.

So there he is, taking aim at the scorpion beside him, shooting his arrow into the dust and gas.

AQUILA

SCUTUM

OPHIUCHUS
& SERPENS

CORNUS

SCORPIUS

CORONA AUSTRALIS

TELESCOPIUM

MAGNITUDE
0 1 2 3 4 5

SCORPIUS
SCO/SCORPII, THE SCORPION

RANK IN SIZE: **33**
ASTERISMS: **THE FISH HOOK**

PERHAPS IT IS NOT so strange that the world sees its gods so differently, when we all look up at such contrarily starred skies. In the northern hemisphere Scorpius, the harbinger of winter and darkness, has always been associated with evil, while the farther south you travel—and the brighter the constellation appears in the sky—so too its malevolence fades. So while the Sumerians knew these stars as Girtab, a poisonous scorpion, and the Egyptians as a serpent, the Bakairi Indians of southern Brazil saw in them a mother carrying a baby on her back. This jagged, stellar S with its distinctive hook may look to you like a pincered arthropod of the sort that killed Orion, but it's all in the eye of the beholder. One man's tail and sting is another man's papoose.

The lucida at Scorpius's center—and so sometimes called Cor Scorpii (The Heart of the Scorpion)—is α Scorpii, a blazing red supergiant 400 times the diameter of the Sun. To the Tūhoe people of New Zealand's North Island, this is Rehua, the chief of all the stars. A deity living in the heights of heaven, Rehua can heal the sick and cure the blind. To Western astronomers, this star is Antares, and—even within the same hemisphere—there's some debate about its name. Deriving from the Greek words *anti* and *ares*, some say it means "the Rival of Ares," and some say it means "Ares's equivalent." Equivocation aside, everyone seeing the world through a Hellenic telescope knows that **Ares** is the Greek god of war, and Mars is his Roman equivalent, and so gets the general gist that this fiery star glitters as pugnaciously as the Red Planet.

There is less violence—no sting in the tale—in the story the Maori tell about these stars. One day the trickster Maui, a descendant of Rehua, went out fishing with the jawbone of his ancestress Muri-ranga-whenua. Having used blood from his nose as bait, he felt something heavy catch at the bottom of the sea. He pulled and he heaved and eventually he trawled up from the depths a massive fish-island covered in grass and trees. Leaving it in the safekeeping of his brothers and villagers, he went to find the holy-man to work out what to do. But when he returned, he saw that they had started hacking into it for food. The fish-island was cleft out of all recognition: its coastline had become serrated and its body full of cliffs, mountains, and valleys. Eventually it split with such force that Maui's fishhook was launched as high as the stars, and that Maui's huge catch split north and south into the islands we now know as the two halves of New Zealand.

SAGITTARIUS

OPHIUCHUS
& SERPENS

LIBRA

LUPUS

RONA AUSTRALIS

NORMA

ARA

TELESCOPIUM

0 1 2 3 4 5

MAGNITUDE

SCULPTOR

SCL/SCULPTORIS, THE SCULPTOR

RANK IN SIZE: **36**
ASTERISMS: **NONE**

LIKE HIS GREEK predecessors, the eighteenth-century French astronomer **Nicolas Louis de Lacaille** clearly considered art a science. Alongside the great instruments of chemistry, physics, mathematics, and astronomy that he honored in his reimagining of the heavens, he placed a painter's easel (Pictor) and a sculptor's chisel (Caelum) and workshop (Sculptor) in the sky.

Unfortunately, since Lacaille first created his *L'atelier du sculpteur* ("the sculptor's workshop"), various features have been chipped off this constellation's block. While he painstakingly fashioned a highly detailed scene on his 1756 planisphere—a laurel-crowned bust on a dignified tripod table, two chisels, and an artist's mallet on an adjoining slab of marble—**Johann Bode** later simplified things. He carved off Lacaille's laurel leaves and the marble block for his 1801 *Uranographia*—though it must be said, he remolded the original bust in exquisite relief. Even the constellation's name was remodeled in the ancient manner to Apparatus Sculptoris, before half of it was lopped off by the English astronomer John Herschel in 1884, leaving just a stump: Sculptor.

It is happily nostalgic though to see the artist's struggle on a celestial par with that of the scientist. Lacaille was born in the midst of a revolution that would eventually see the world transformed to a place of empiricism, rationality, medicine, and progress, but a world where scientific lust soon began to patronize the primitive, emotional, unprovable realm of art; where the ability to name and number every last star might sometimes rob the sky of its mystery.

So, despite his predilection for placing scientific instruments in the sky, it is heartening to remember that in many ways Lacaille himself—like all astronomers before and since—was an artist: chiseling images out of otherwise random collections of stars, sculpting sense out of the firmament.

CETUS

AX

GRUS

PHOENIX

IDANUS

0 1 2 3 4 5

MAGNITUDE

SCUTUM

SCT/SCUTI, THE SHIELD

RANK IN SIZE: **84**
ASTERISMS: **NONE**

HE WAS SIMPLY not the man he had once been. As the horses led the coach out of the city gates, **Johannes Hevelius** berated himself in frustration. He had powered through life with a grueling work ethic akin to religious fervor. Despite being obliged to take over the family brewery and working tirelessly as a magistrate for the city council, he had made his own astronomical instruments, ground the very glass for his telescopes, and still found time to construct the finest observatory in the world—on top of his own house. But his sixty-eight years were beginning to take their toll. He had woken up in the middle of the night plagued by an unknown fear; and now here he was, riding out of Danzig to their countryside retreat—to recuperate! The disappointment sat like metal in his mouth. He avoided **Elisabetha**'s gaze.

His young wife looked out of the coach window. It was a clear September night. She was sad not to be in the observatory, but she scanned the heavens with a trained eye and settled on a dark patch to search for faint stars. (Eleven years later in 1690, she would introduce this segment of the night sky to the world as the constellation Vulpecula with the publication of her dead husband's star atlas.) By the time they arrived, a southerly wind had picked up and the sky had clouded over. They sent the coachman back home and went straight to bed.

They were asleep then when the coachman drove back through the city gates just before they closed. They were asleep when he returned the horses to the stable. And they were asleep when, either by accident or malice (Hevelius always suspected the latter), the coachman left a candle burning in that same stable that set it alight. The flames licked through the wood and the straw, burned the family's fine horses alive, and at last caught the side of the house. Fanned by a strong wind from the south, the fire ravaged Johannes and Elisabetha's home, destroying almost all of their instruments, manuscripts, books, and household possessions. Despite the best efforts of valiant neighbors who threw all they could out of the windows and rescued several astronomical tomes, most of Hevelius's work, and moreover his entire observatory, was destroyed.

When the patronage of King John Sobieski III of Poland enabled the bereaved astronomer to restore his world-famous observatory to its former glories, there was of course only one way to thank him. In 1684, Hevelius placed Scutum, "Sobieski's Shield," in the sky.

SAGITTA

HERCULES

AQUILA

OPHIUCHUS
& SERPENS

SAGITTARIUS

0   1   2   3   4   5

MAGNITUDE

SEXTANS

SEX/SEXTANTIS, THE SEXTANT

RANK IN SIZE: **47**
ASTERISMS: **NONE**

SIX ILLUMINATING FACTS ABOUT THE SEXTANT IN THE SKY

1   Not the same as a nautical sextant. That instrument measures angles between heavenly bodies and the horizon; this one measures the positions of the stars—although both are 60° devices, i.e., a sixth ("sextant" in Latin) of a circle.

2   Not the same as the grave digger in *Hamlet*. That's a sexton.

3   Three of its stars form the constellation Tianxiang, the celestial prime minister of Chinese astronomy.

4   It commemorates the brass instrument owned by **Johannes Hevelius**—also the constellation's inventor—that was destroyed in the 1679 fire at his famous observatory in Danzig. (There is a very touching illustration of him and his wife **Elisabetha** using it together.)

5   It does not commemorate the 118-feet-wide sextant constructed by the Timurid ruler and astronomer Ulugh Beg for his observatory in Samarkand, Uzbekistan (which was built in the 1420s, destroyed in 1449, and, despite having been one of the greatest observatories in the Islamic world, only rediscovered in 1908).

6   It was originally called Sextans Uraniae, meaning "the Sextant of **Urania**," after the Greek muse of astronomy.

0 1 2 3 4 5

MAGNITUDE

TAURUS

TAU/TAURI, THE BULL

RANK IN SIZE: **17**
ASTERISMS: **THE HEAVENLY G, THE HYADES, THE PLEIADES, THE V,
THE WINTER OCTAGON, THE WINTER OVAL**

TWO BOVINE tales of lust ignite the bright-starred horns of Taurus. Both—you will not by now be surprised to hear—involve the libidinous **Zeus**.

The first is of **Io**: yet another of the god's flings flung aside to a life of suffering and misery. The story starts much as they all do: once upon a time there was a beautiful girl and Zeus fell in love with her and (to use the euphemism the ancients seem fond of) ravished the innocent virgin. When the god's wife, **Hera**, challenged her adulterous husband, he lied to her face: "I did not so much as touch the girl," he said brazenly, though not quite able to look at her straight. Hoping to protect young Io, who still softened his capricious heart, Zeus turned her into a cow; but Hera claimed the heifer as hers and sent the hundred-eyed guardsman **Argus** to spy on her. Not one to be trumped in marital guile, Zeus sent **Hermes** to steal her back. Charming Argus asleep with the soft sounds of a flute, Hermes cut off his head and freed poor lowing Io. In revenge, Hera send a gadfly to chase, bite, buzz, and plague Io wherever she went while Argus's hundreds of eyes she placed forever in a peacock's feathers.

The second is the rape of Europa: a bovine tale in reverse—for when Zeus was not turning his paramours into heifers, he was disguising himself as a bull. Europa was another young and lovely thing, the daughter of King Agenor of Phoenicia. Zeus looked down on this princess lasciviously—or perhaps in this case, his heart was tender and true—as she played on the beach with her friends, her delicate ankles frolicking in the surf. Employing the wiles of his trickster son Hermes once again, Zeus devised a cunning plan. King Agenor had a herd of fine cattle, and getting Hermes to drive them from their mountain pastures to the seashore, Zeus took his place among the line of cows plodding their way down the hillside. Europa turned from the waves to see a beautiful white bull staring up at her, batting thick lashes across its big brown eyes. She stroked its back, and let it lick her palm, and put flowers in its horns. Then climbing up onto its back, she let it carry her down to the shore. But when it reached the water it didn't stop, and soon it was swimming with her out to sea. Europa's friends stood in horror as the bull diminished toward the horizon, carrying the princess to Crete and leaving a trail of flowers on the waves.

AURIGA

PERSEUS

ARIES

ORION

ERIDANUS

LEPUS

MAGNITUDE

0 1 2 3 4 5

TELESCOPIUM

TEL/TELESCOPII, THE TELESCOPE

RANK IN SIZE: *57*

ASTERISMS: **NONE**

IT'S FREEZING and I've been standing here for ages. Even though this is my present—you gave it to me—you seem already to have commandeered it.

"Hang on. I think I've got it now. I think maybe it's the other way around. This bit goes here."

You fiddle with one of the lenses again, unscrewing something and attaching an indeterminate piece of plastic in its place. I shift my weight onto the other foot and bury my hands farther down into my pockets. It's astounding how clear the stars are, here in the east of this polluted city, under a cloudless Monday night.

"It's still just completely out of focus."

You're getting grumpy now. I don't try, as I did unsuccessfully about ten minutes ago, to intervene.

When **Galileo**'s contemporaries first looked through his telescope, many of them denied that they could see anything at all. Perhaps they weren't lying. Perhaps the clergy objected to the wobbly, tunnel-visioned instrument (far less sophisticated even than the one you are currently trying to construct) not because it supported the revolutionary assertions of **Copernicus** and **Kepler** and challenged religious hegemony, but rather because it was so annoying to use. Of course Galileo himself did not invent the telescope that **Lacaille** later honored as a constellation. It was first created by a Dutch spectacle-maker called Hans Lipperhey, who in September 1608 applied to the States General in the Hague for a patent for an invention by which "all things at a great distance can be seen as if they were nearby."

"OH MY GOD!"

You are leaping about, whooping. Finally, you let me look. And there it is. The Moon.

The Moon, which until this very moment has been to me a vast and empty metaphor. Something that haunted the dark spaces of the mind of writer Sylvia Plath; or that the depressive Philip Larkin stared at glumly through his curtains in his poem "Sad Steps," which directly inverts one of poetry's seminal works of stargazing, Philip Sidney's 1591 *Astrophil and Stella*.

The Moon, which is cratered in white and gray, and moving so fast that I actually *see* it journey across the lens and vanish out of sight. The Moon that men have stepped on, but which until now made no sense to me. Poet Michael Donaghy's "bright disc shining in the black lagoon, perceived by astrophysicist and lover," that I never really believed in, that I now witness orbiting the Earth.

You show me the Moon. In the present tense. This is no symbol, but the greatest expression of love.

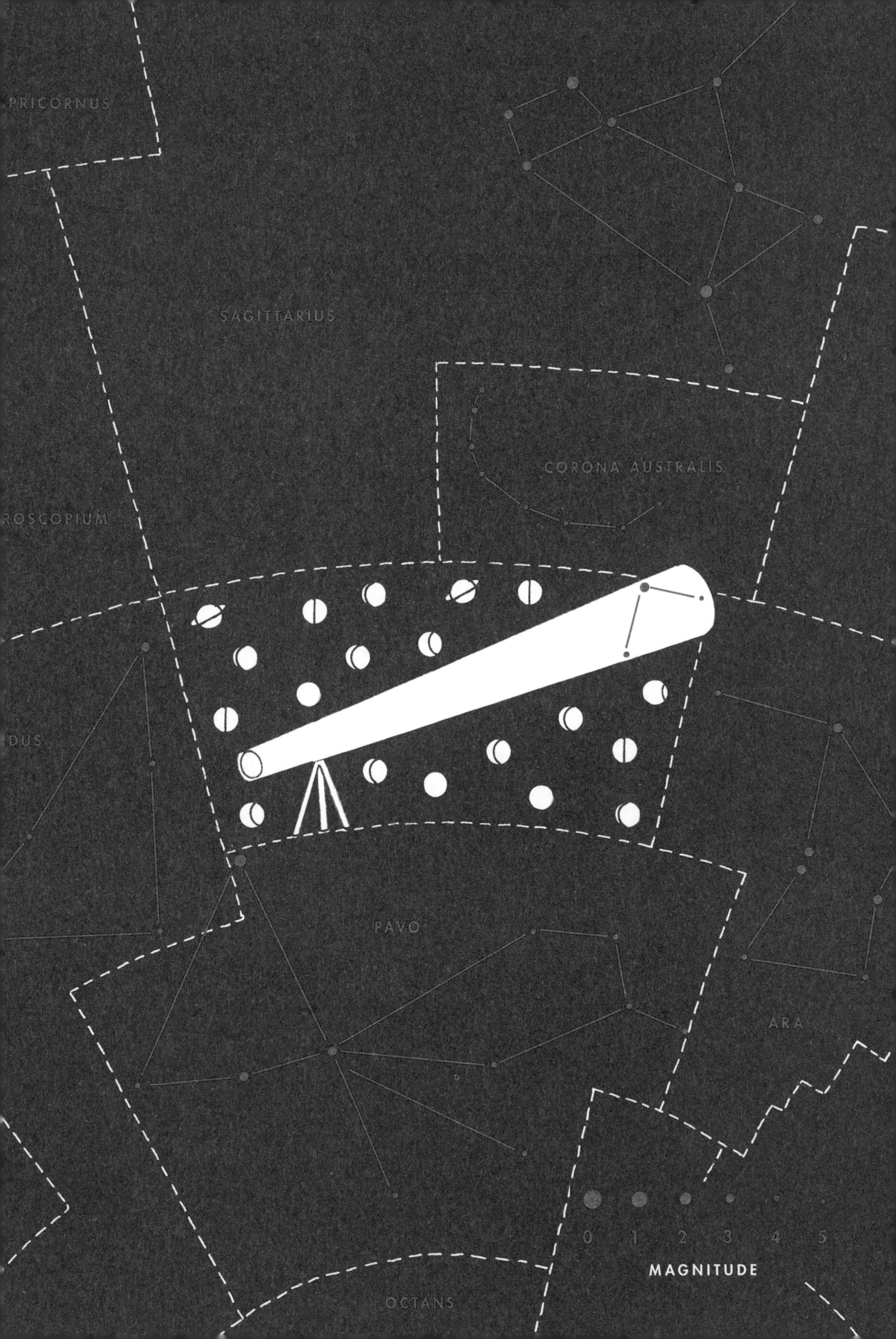
PRICORNUS
SAGITTARIUS
CORONA AUSTRALIS
ROSCOPIUM
DUS
PAVO
ARA
OCTANS
0 1 2 3 4 5
MAGNITUDE

TRIANGULUM

TRI/TRIANGULI, THE TRIANGLE

RANK IN SIZE: 78

ASTERISMS: NONE

FOR THOSE OF US who spent long and miserable hours of our lives in examination rooms struggling to calculate the mathematical properties of an isosceles triangle, drips of sweat dropping from our panicked brows and onto the irritatingly small desk in front of us—which more often than not wobbled vindictively on one leg—the resultant grade was often a dreaded Delta. For the mathematically blessed ancient Greeks, however, delta was the fourth letter of their alphabet, the numerical value of which was four, and the uppercase of which was an isosceles triangle: Δ. (Little delta looks like half a pair of glasses from a side-on angle; or a bottom-heavy and incomplete number 8; or, more pertinently, quite a lot like our little d, but written by someone who's had one too many drinks: δ.) It has been used ever since to designate things as diverse as scientific concepts and indie bands: to music fans, Δ signifies the Mercury Prize–winning band Alt-J, so named because that is the symbol you get if you type those keys on an OS X Apple keyboard. To less materialistic musicians, it is a major seventh jazz chord.

The Greeks derived their Δ from the Phoenician letter *dalet*, also a triangle, which they rotated and skewed the angles of in a calculation that would have sent shivers down your spine, had you been asked to solve it in the cold, clock-ticking gym hall. Of course the Grecian polymath Eratosthenes could probably have worked the sum out in his sleep. That Hellenic mathematician–poet–musician–astronomer who invented geography, and was the first man to calculate—with remarkable accuracy—the circumference of the Earth, and the tilt of its axis, was the same man who looked up into the night sky and decided that the three bright stars of this constellation the ancients called Deltoton (because of its resemblance to Δ)—represented the delta of the river Nile.

To the Romans the constellation was associated with Sicily, the triangle-shaped island once called Trinacria—or so the mysterious Latin author Hyginus tells us. Much of our star lore comes from Hyginus's didactic poem, the *Poeticon astronomicon*, a tricky tome first officially published in Venice in 1482. Though rich in celestial stories, its Latin is so poor that some historians have assumed that it is in fact the notes of a schoolboy recapitulating another work. Nevertheless, whoever Hyginus really was, he tells us that the Romans saw this constellation as Sicilia, the island sacred to Ceres, from which her daughter Proserpina was stolen by Pluto: a teleological tale to explain the seasons that the Greeks saw in the stars of Virgo.

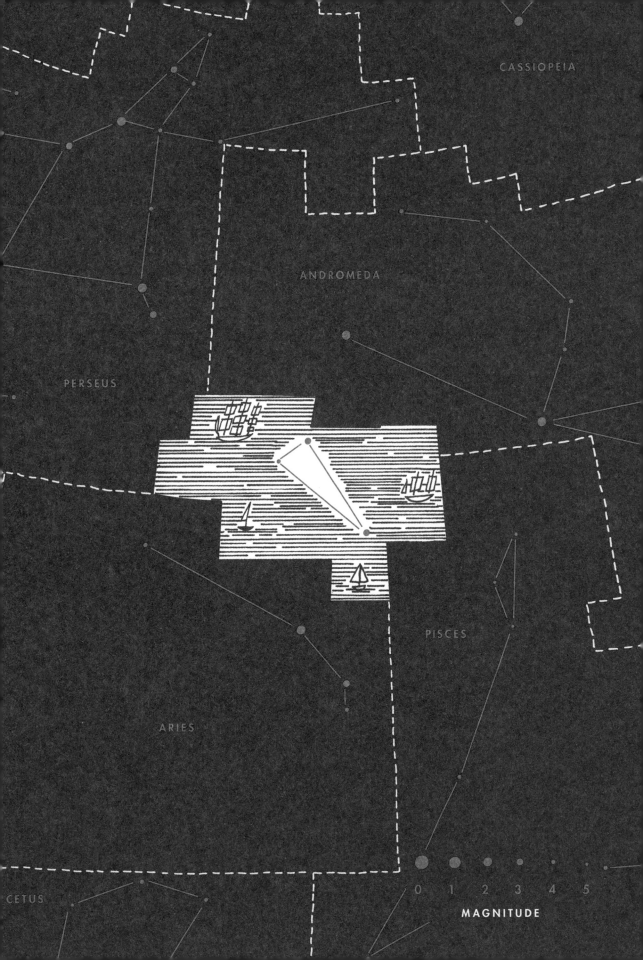

CASSIOPEIA

ANDROMEDA

PERSEUS

PISCES

ARIES

CETUS

TRIANGULUM AUSTRALE
TRA/TRIANGULI AUSTRALIS,
THE SOUTHERN TRIANGLE

RANK IN SIZE: **83**
ASTERISMS: **THE THREE PATRIARCHS**

NOT IN FACT a schoolboy innuendo, the Southern Triangle is one of a trio of building and surveying instruments alongside Norma and Circinus in the southern skies. While the latter two are inventions of the eighteenth-century French astronomer **Lacaille**, Triangulum Australe featured on a star atlas much earlier, in 1603— though it was first acknowledged by Italian navigator Amerigo Vespucci exactly a hundred years before even that. It is, however, the Dutch navigators **Keyser** and **de Houtman** who are usually credited with its conception. They passed on their observations from their southern sea voyages to the astronomer **Petrus Plancius**, whose 1589 globe was used to create the Uranometria atlas of **Johann Bayer** on which it first appeared.

To add yet another layer of complication, Lacaille's planisphere of 1756 labeled it as "*le Triangle Austral ou le Niveau*" ("the Southern Triangle or the Level"), but as Ian Ridpath—the venerable astronomer without whose books I would have been utterly unable to make heads or tails of the brain-dizzying topography of the stars—tells us: "Through some misreading, the historian R. H. Allen transferred the appellation 'level' to the nearby constellation Norma and termed that constellation the Level and Square (instead of the Rule and Square), thereby confusing generations of astronomers."

All this etiology is, however, slightly irrelevant to the generations of navigators who have sailed their ships under the three shining corners of this triangle (the stars of which are more luminous, though smaller, than those of its northern counterpart, Triangulum). Utterly unable to be seen north of the tropics, this bright constellation is essential to orientation in the skies of the southern hemisphere and was especially so before the days of GPS and satellite navigation; or indeed the marine chronometer that enabled sailors to determine their longitude (the story of which is told in the celestial timekeeper Horologium). Intrepid explorers seeking new lands would look up to its lights in relief; migrating birds crossed continents under its sway. While European sailors tried to harness this stellar power with astrolabes and octants, the indigenous inhabitants of the lands they sought to conquer accurately sailed across the oceans with no other instruments than their cerebra. Using the mnemonic power of song to pass down ancient navigational wisdom, Polynesians paddled their canoes according to the rising and setting of stars, such as those of Triangulum Australe, whose movements were inscribed in their consciousness. Three is indeed a magic number.

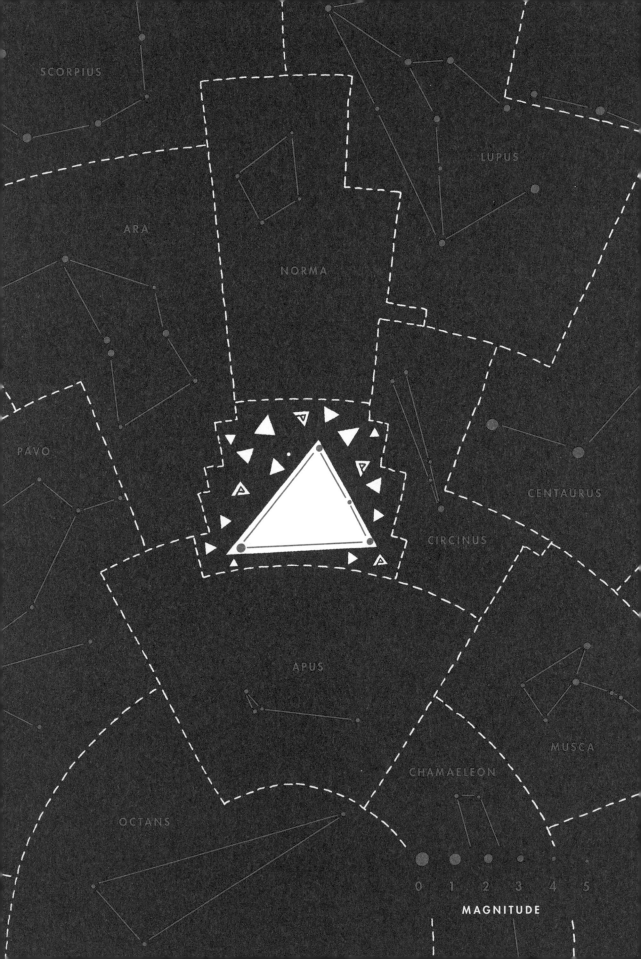

SCORPIUS

LUPUS

ARA

NORMA

PAVO

CENTAURUS

CIRCINUS

APUS

MUSCA

CHAMAELEON

OCTANS

0 1 2 3 4 5

MAGNITUDE

TUCANA
TUC/TUCANAE, THE TOUCAN

RANK IN SIZE: **48**
ASTERISMS: **NONE**

THE TUPI PEOPLE of Brazil had been living with this tropical bird for thousands of years before Portuguese explorers first clapped eyes on it, as they stepped onto the shores of a strange new world and stared in astonishment at the rainbow-billed creature the Tupi called *tukana*. Just as the South Sea islanders had been telling their children legends about the stars long before the intrepid Dutch navigators **Keyser** and **de Houtman** officially introduced this constellation in the late sixteenth century. Yet it was **Petrus Plancius**, our flat-footed astronomer, who decided to put the bird and the stars together in astronomical posterity and give the constellation its name.

There is no story, therefore, about a kindly toucan who carries a baby to safety in its beak to explain the bird's presence in the firmament. There is no dark nursery tale about the boy who poked his nose into a toucan's nest—they make their homes in tree hollows—and was pecked to painful death. But nestled in the depths of Tucana, light-years away, are two of the most famous deep-sky objects in space—the Small Magellanic Cloud and the globular star cluster 47 Tucanae—and there is a gruesome story behind the man whose name has been given to the first of these.

Ferdinand Magellan was born in 1480, lost his parents when he was ten, and was injured in service in Morocco, leaving him with a lifelong limp. Yet this Portuguese explorer was a brave navigator who set sail on the scent of the Spice Islands, and in trying to find a westward route to these lucrative islands in the east—the Moluccas, in modern-day Indonesia—he organized the expedition that completed the first circumnavigation of the globe in 1522. Magellan was the first European to write home about a hazy patch of light in the southern sky, about seven times the diameter of a full moon, that looks as though it has broken off the Milky Way but is in fact a dwarf galaxy full of radiation, stellar winds, and clusters of brilliant, newly born stars. Unfortunately, however, he did not live to see the ship's full circle home. The Portuguese had been colonizing and Christianizing the indigenous tribes they came across, but Lapu-Lapu, the leader of Mactan in the Philippines, was not so willing to be conquered. When Magellan's forces arrived on the chieftain's shores, they were besieged by unprecedented force. Magellan himself was struck by a bamboo spear, and when the islanders realized who he was, they turned on him in fury. As his defeated crew retreated to their ships in despair, they saw their captain staring out to them at sea as he was torn apart by men, cutlasses, iron, and rage.

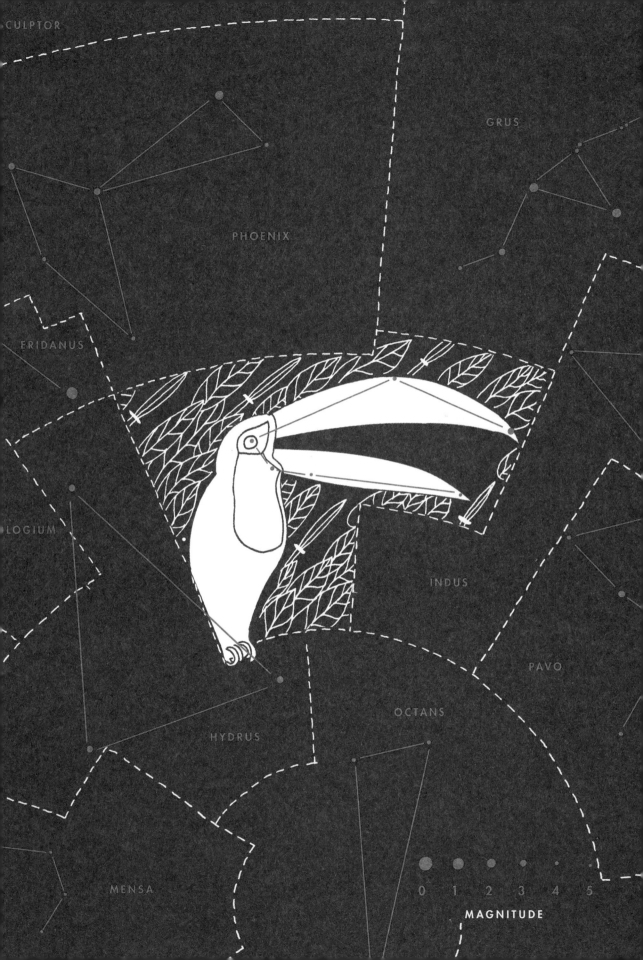

CULPTOR

GRUS

PHOENIX

ERIDANUS

LOGIUM

INDUS

PAVO

OCTANS

HYDRUS

MENSA

0 1 2 3 4 5

MAGNITUDE

URSA MAJOR

UMA/URSAE MAJORIS, THE GREAT BEAR

RANK IN SIZE: 3
ASTERISMS: THE ARC, THE BIER, THE BIG DIPPER, THE HORSE AND RIDER,
THE POINTERS

THIS IS IT. The one you know. The thing that you might actually recognize in the night sky. The Big Dipper, the Saucepan, the Plow: call it what you like, the famous star-pattern (or asterism) at the heart of Ursa Major is one of the very few that actually looks like the things it's meant to resemble. This is the stargazer's starting block. For once you have delightedly connected the dots of the seven bright stars to form a celestial ladle, you can then easily follow the line of the ladle's far edge—from the star Merak to Dubhe—straight up to **Polaris** in Ursa Minor. This is the North Star, our current Pole Star: the star almost directly above the North Pole around which the whole sky seems to rotate, the epicenter of the magic lantern of myths that cast their spell over the night sky.

All this of course is if you live in the northern hemisphere. If you live in the southern hemisphere, things are *quite* different. Ursa Major is not so major there, where from 40° south it is only partly visible, and at middle latitudes it is lost entirely. Western powers made their astronomy, like so many things, the official way of seeing. Although interestingly Ursa Major was also seen as a bear in Native American and Hebrew traditions, and the Druids called the Big Dipper "Arthur's plow." But for now, let's not worry about anthropological politics. Let's kick off our slippers, sit back, and enjoy the story. Are you ready? Then I'll begin.

Zeus of course was king of the gods: lord and goddest of all. His sister, **Hera**, was also his wife, and the children he fathered left, right, and center were many. One of his illegitimate daughters was **Artemis**—the goddess of the hunt, and of lions and stags and all things wild. Now Artemis would range through the forests and the plains with her bow and arrows, and a loyal band of nymphs who had sworn a solemn vow of chastity to the goddess; and one of those feral virgins was **Callisto.**

One day, as this pretty little wood nymph was out picking a fine branch to hew into an arrow, she caught the eye of Zeus. Disguising himself as Artemis, he approached the naive Callisto. She turned and saw her leader and smiled. But before Callisto knew what was happening, Artemis was suddenly Zeus, the branch in her palm had slipped from her grip, and her innocence and chastity had been slashed and burned in one fell swoop.

Whether she bore the god's child straight away, or gestated in the woods for the more quotidian nine months, we can't be sure. What we do know is that their offspring was a boy called **Arcas**, and that he was brought up by his grandfather **Lycaon**, a very nasty man. You can read all about that in Lupus, but this is the story of Ursa Major; and currently Callisto is roaming alone through the forests as precisely that: a great bear. For that is what Zeus's jealous wife, Hera, turned her into in a rage: she made thick brown hairs sprout from Callisto's fine skin, and her slender spine and limbs curve over and squat into an ursine stoop.

The years passed—fourteen, perhaps fifteen—and it was a bright autumn morning and light streamed down through the canopy. Hearing some unusually heavy rustling in the leaves, Callisto turned to see a teenage huntsman aiming his arrow directly at her. Immediately, and with a mother's love, she knew it was her son. As she cried and wept and called out his name, Arcas of course heard only the loud growls of a bear. Terrified of the raging creature, he bent back his bow.

To be continued . . .

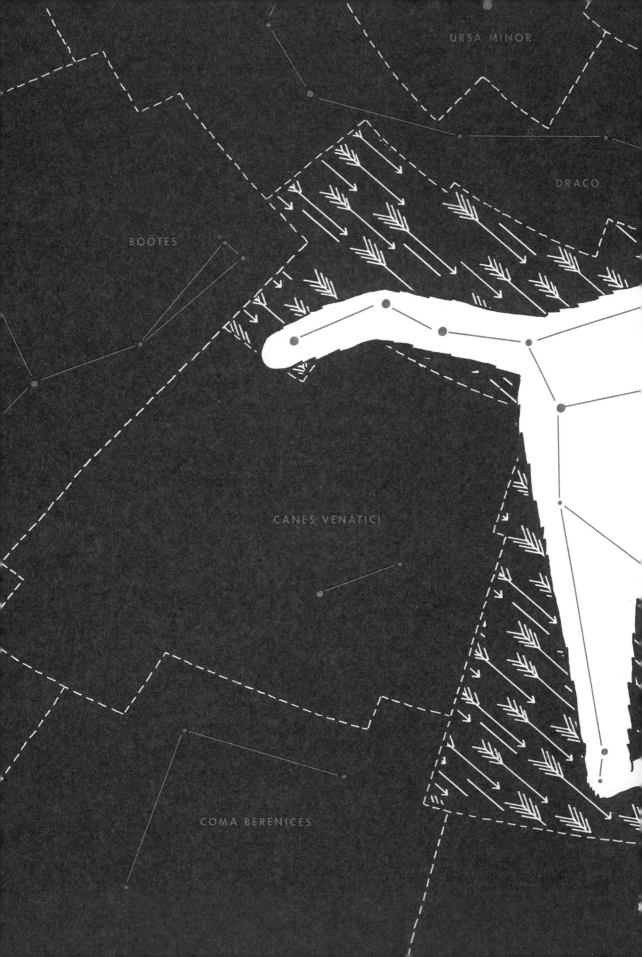

URSA MINOR

DRACO

BOÖTES

CANES VENATICI

COMA BERENICES

CAMELOPARDALIS

AURIGA

LYNX

GEMINI

CANCER

LEO MINOR

LEO

● ● ● ● · ·
0 1 2 3 4 5

MAGNITUDE

URSA MINOR
UMI/URSAE MINORIS, THE BEAR CUB

RANK IN SIZE: **56**
ASTERISMS: **THE GUARDIANS OF THE POLE, THE LITTLE DIPPER**

THE GREEKS navigated using Ursa Major, but for the Phoenicians, Ursa Minor was the constellation by which to travel the seas. Just like its more famous counterpart, you can find another saucepan in the flanks of this little bear: a smaller plow-shape diminutively known as the Little Dipper, whose handle runs the line of its tail. And right at the tip of this elongated tail is the star that this northernmost constellation is famous for—α Ursae Minoris, the Pole Star. This bright star lies within half a degree of the Celestial North Pole. The Celestial Sphere is an imaginary globe projected from the Earth's surface into the sky; so the Celestial North Pole is the zenith—that is, the point on the Celestial Sphere directly overhead—for someone looking up from the Earth's North Pole. But Polaris, as α Ursae Minoris is known, has not always been our Pole Star—the point that stands almost perfectly still as the heavens spin around it. When **Ptolemy** was mapping the stars in the second century AD, it was in fact 11° away. This is because of something called precession—the slow but incremental process by which the Earth wobbles on its rotational axis. About once every 26,000 years the axis completes one precessional cycle, tracing out the shape of a cone. Which all means that the position of the stars moves by about 1° every seventy-two years.

But why am I harping on about an astronomical phenomenon? You have just been reading the story of Ursa Major and are hanging on its cliff edge—with poor **Callisto**, now a bear, in the middle of the woods and her unsuspecting son **Arcas** pointing his arrow in her direction.

Luckily, and as fate would have it, the gods intervened. Zeus got wind of the ursine matricide about to occur, and, in the divine nick of time, he turned the young Arcas into a bear as well so that he could understand his mother's cries. Then, so as not to be scratched by their sharp claws, he picked them both up by their tails and spun them around and around and lassoed them into the sky. (That is why, incidentally, unlike ordinary bears, Ursa Major and Ursa Minor have such long tails.) But ever-jealous **Hera** was not satisfied with this. Traveling to the underwater palace of her brother **Poseidon**, king of the sea, she persuaded him never to let the bears enjoy bathing in the heavenly waters. Poseidon granted his sister her vindictive wish, and this is why, for most of the northern hemisphere (where the story began), the stars of Ursa Major and Ursa Minor will never dip below the horizon. Conversely, for those stargazers living south of Rio de Janeiro or Alice Springs, these two bears will never rise.

CASSIOPEIA

CEPHEUS

DRACO

CAMELOPARDALIS

URSA MAJOR

ÖTES

0 1 2 3 4 5

MAGNITUDE

VELA

VEL/VELORUM, THE SAIL (OF ARGO NAVIS)

RANK IN SIZE: **32**
ASTERISMS: **NONE**

JASON AND THE ARGONAUTS: AN EPIC TALE TOLD BACKWARD
(in which fifty Hellenic heroes, etc.)

PART III: VELA

ZEUS'S PET RAM carries **Phrixus** and **Helle** in flight across the skies. It has swept the young children of King Athamas from the altar on which they were about to be sacrificed—thanks to their evil stepmother Ino's dark machinations—and now carries them east, to Colchis. The night is cold and the skies loom large and Phrixus digs his nails deep into the ram's golden fleece. But Helle starts to lose her grip. Phrixus feels his sister's delicate arms slipping from his waist until, turning around in panic, he sees little Helle fall through the sky to the sea: lost forever in the narrow depths of the Dardanelles, the strait the Greeks now call Hellespont. Arriving at Colchis, on the shores of the Black Sea, the grieving boy sacrifices the ram to **Zeus** in thanks for his salvation and gives its dazzling fleece to the ruler of that land. **King Aietes** hangs it in an oak tree in a sacred grove of **Ares**, the god of war, and leaves a sleepless dragon to keep it guard.

Years later and back in Iolcus—the homeland from which Helle and the now dead Phrixus once fled—a man in one sandal arrives at the door of Pelias. Some time ago, Pelias had usurped the throne from its rightful owner, his half brother Aison. Terrified that their newborn son **Diomedes** would be murdered, Aison and his wife secreted the babe-in-arms in the hills of Mount Helicon, to be brought up by that famed raiser of heroes, wise **Chiron** the centaur. Pelias ruled on, oblivious, perturbed only by an oracle that told him to beware of a man in one sandal, for this would be the man who dealt his death.

Jason, for that is the name of this uni-shod young man standing before the terror-struck king, has no idea about this prophecy. What he does know is that Jason was not the name he was given at birth, and that he is the rightful heir to Pelias's throne. He has come here precisely to reclaim it. Though afraid to deny him his birthright, his usurping half uncle plays a slippery trick. The throne will be Jason's, of course, but first he must go to Colchis and bring back the troubled ghost of Phrixus that is haunting Iolcus, along with the Golden Fleece—a task so arduous the king has no fear that Jason will succeed.

But standing before Pelias is a man with steel in his sinews: a hurt soul tutored in a hero's ways. Not two weeks later, he has hewn a ship with the help of **Athene**, and Jason and the Argonauts hoist their sails.

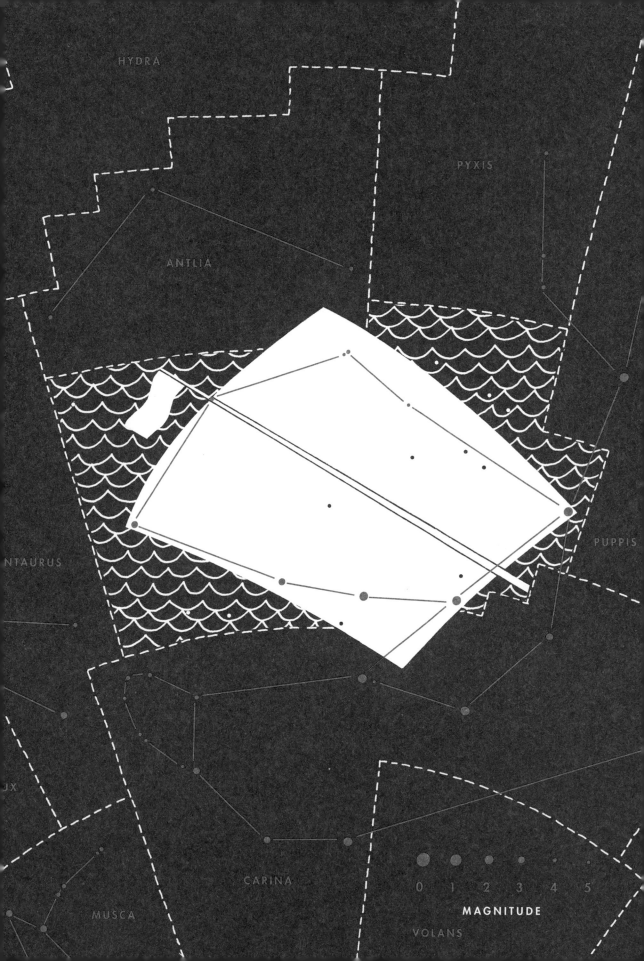

HYDRA

PYXIS

ANTLIA

PUPPIS

NTAURUS

JX

CARINA

MUSCA

VOLANS

0   1   2   3   4   5

MAGNITUDE

VIRGO
VIR/VIRGINIS, THE MAIDEN

RANK IN SIZE: **2**
ASTERISMS: **THE DIAMOND, THE SPRING TRIANGLE, THE Y**

FIRST THERE is that awakening deep in the marrow of your winter-shrunk bones, that quality of hope in the light, and then there are daffodils, and evenings. Evenings! You'd forgotten that days didn't turn straight into night before you'd even left work. Spring! Spring has sprung! Next, these heart-speeding spring days mellow into early summer, and the whole world hovers on a heady, glorious axis: things are no longer newborn, but are not yet fully grown; the Sun is warm, but the air is still cool; everywhere is blossom and bloom, and everything is promise, but the crops are not yet cut. Across centuries and continents humans have known this as the time of the virgin—fertile, plump, and ripe for the plucking. So of course this is the season when, if you look up to the stars, you will see one in the sky. Holding an ear of wheat, or a sheaf of corn, and held aloft in the night sky by angelic wings is Virgo.

The Babylonians called her Ishtar. Or Ashtoreth. Or Astarte. The Egyptians associated her with Isis, their goddess of magic, motherhood, and fertility. Virgo's lights appear most brightly in the sky just at this heady, liminal time of year, when Anglo-Saxon pagans would have been offering colored eggs to their own goddess of fertility and spring, Eostre, placing these painted talismans on graves in a festival of rebirth that gave us the Christian Easter. Although it is the second-largest constellation in the sky, Virgo's stars are fairly faint. Except, that is, for the brilliant first-magnitude Spica—a blue or blue-white lucida that, as its Latin name tells you, marks the ear of grain she holds in her hand. Desert Arabs sometimes called this lonely-seeming star Azimech, from *al-simāk al-a'zal*, meaning "the Undefended." It may look solitary to the naked eye, but it is easy to find; especially with the help of the old stargazer's adage: *follow the Arc to Arcturus and then speed on to Spica*. That is to say, trace a line through the handle of the Big Dipper in Ursa Major and continue southward through Arcturus in Boötes and then down, and you will find α Virginis, or the Virgin's Spike as it was known in Old England.

But we are not in Albion. We are in a meadow in Sicily, and Persephone is skimming her fingers over the long grass. Occasionally she picks a flower. Occasionally the breeze catches a strand of her hair. Bending down to smell a narcissus in bloom, she feels the earth crack beneath her feet and she is sucked all the way to Hades. She doesn't return that evening, or the next, and her mother, **Demeter**, searches for her furiously, racked by worry. Since Demeter is the goddess of grain, her grief is felt in the fields. The crops wither. She asks the Great Bear if she saw anything in the night, but only the sun god **Helios** witnessed what happened. When Demeter hears that **Hades**, the king of the underworld, has stolen his own niece—and taken the virgin as his spoil—she fumes her way to Persephone's father, **Zeus**, and forces him to rescue her.

The cunning Hades, however, now plays a cruel trick. Like all good girls, Persephone knows that if you are ever abducted and taken to the underworld, you must not touch a morsel of food or you can never again reside in the land of the living. And not one crumb has thus far touched the reluctant bride's lips. But as news reaches her and her uncle that she is about to be freed, Hades prepares her a farewell feast. Not when he offers her the finest loaves or the sweetest wine is she tempted, until finally he hands her a pomegranate. She can't resist and has just six of its succulent seeds. And with those six small transgressions, she seals her fate. She will have to spend a third of every year in the underworld, and as winter comes the Earth will mourn and the crops will die, until she returns each year, in the spring.

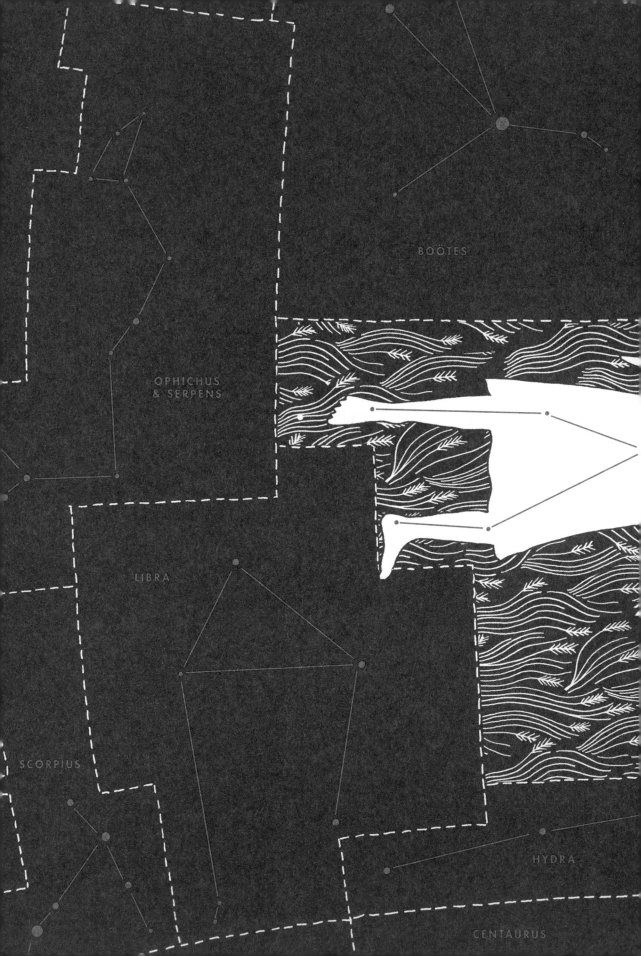

BOÖTES

OPHICHUS
& SERPENS

LIBRA

SCORPIUS

HYDRA

CENTAURUS

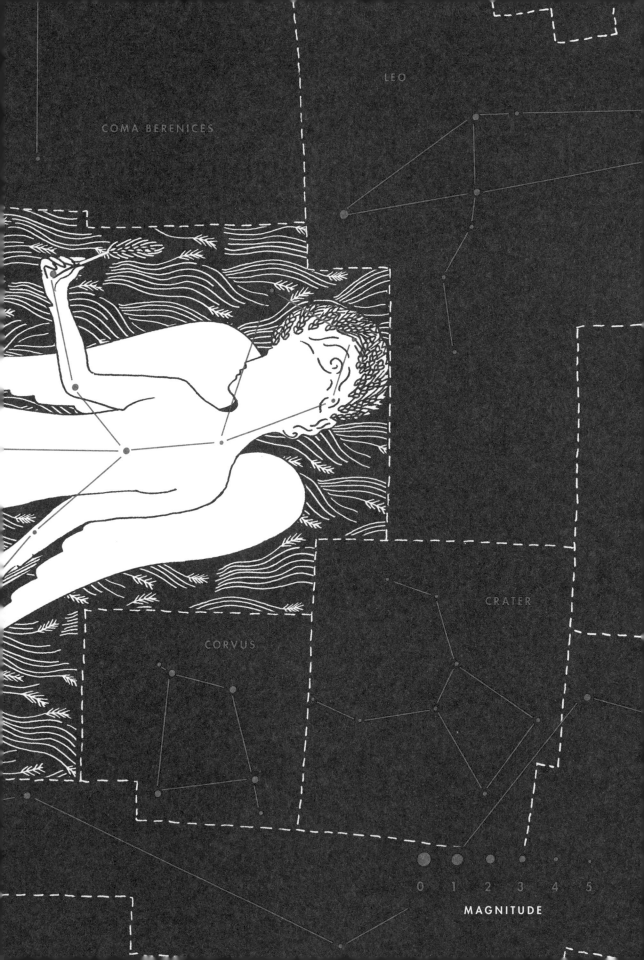

COMA BERENICES

LEO

CRATER

CORVUS

0 1 2 3 4 5

MAGNITUDE

VOLANS

VOL/VOLANTIS, THE FLYING FISH

RANK IN SIZE: **76**
ASTERISMS: **NONE**

A HAIKU about a brief moment toward the end of a long day at the close of the sixteenth century when the weary explorer **Pieter Dirkszoon Keyser** looked out of his ship and saw a flying fish skimming over a hitherto unsailed patch of tropical sea.

Fin-winged, sail-flying
Astonished fish and sailor
Greet the alien's gaze.

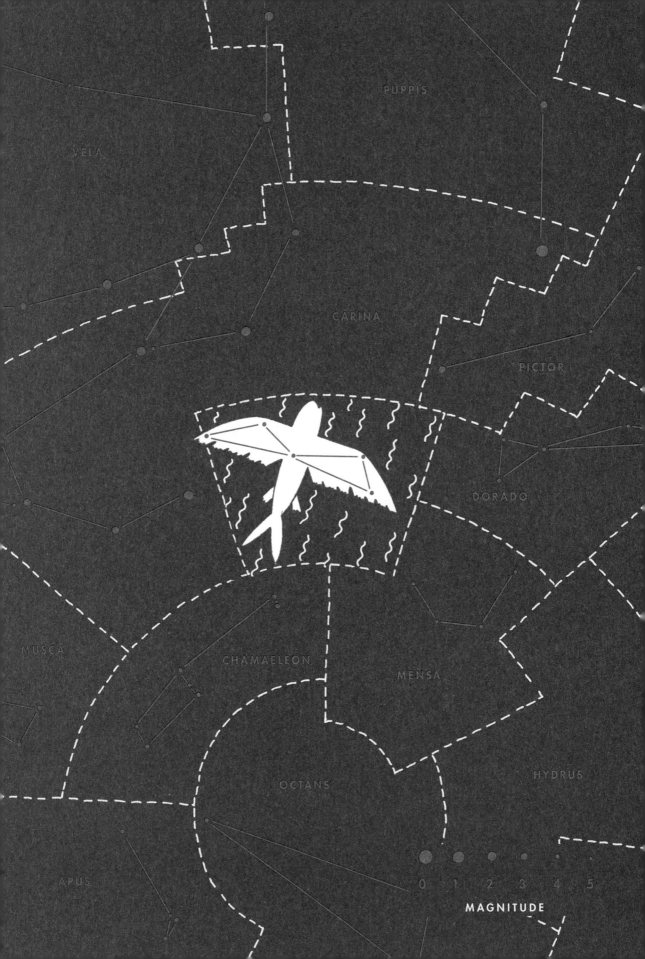

PUPPIS

VELA

CARINA

PICTOR

DORADO

MUSCA

CHAMAELEON

MENSA

HYDRUS

OCTANS

APUS

0 1 2 3 4 5

MAGNITUDE

VULPECULA

VUL/VULPECULAE, THE LITTLE FOX

RANK IN SIZE: **55**
ASTERISMS: **NONE**

WHEN MY FATHER told me the story it was a fox, a hen, and a bag of corn. In the early medieval math book *Propositiones ad Acuendos Juvenes* ("Conundrums to Sharpen Young Minds"), it was a wolf, a sheep, and a cabbage.

Of course, as I sat on my father's knee when he came home from work and told me a bedtime story—the smell of whisky always sits me back there—and listened again to the riddle I knew so well (but could never remember how to answer), I had no idea that it was known as a "river-crossing puzzle." Or that it goes back at least as far as the eighth century and has teased minds from Romania to Zimbabwe to Scotland to Cameroon.

...

Once upon a time, there was a farmer who went to market and bought a fox, a goose, and a bag of beans. He was very happy with himself and how big his wife's smile would be when she saw the plump goose, the bushy fox, and the overspilling bag of beans.

He took them to the river so that he could row them across it on his little blue boat and wend his way home. But as he stepped onto his boat, he soon realized that it was only big enough to carry either the goose, the fox, or the bag of beans with him across the river. He would have to leave the other two on the shore and come back for them.

But if he left the goose with the fox, the fox would eat the goose, and if he left the goose with the beans, the goose would eat them too! The unhappy farmer scratched his head.

How could he possibly carry his wares home safely to his hungry wife?

...

History says there are no legends associated with this constellation, but perhaps when, in 1687, the Polish astronomer **Johannes Hevelius** named his new constellation "Vulpecula cum Ansere" (The Little Fox with a Goose), he had this folklore puzzle in mind.

Just as I always did when I sat on my father's knee, astronomers over the centuries seem to have forgotten how to solve the riddle of this story and the goose has now disappeared from the constellation—perhaps the cunning fox has eaten it. Although some say the poor bird is hiding in the constellation's brightest star, Alpha Vulpeculae, known to some as Anser.

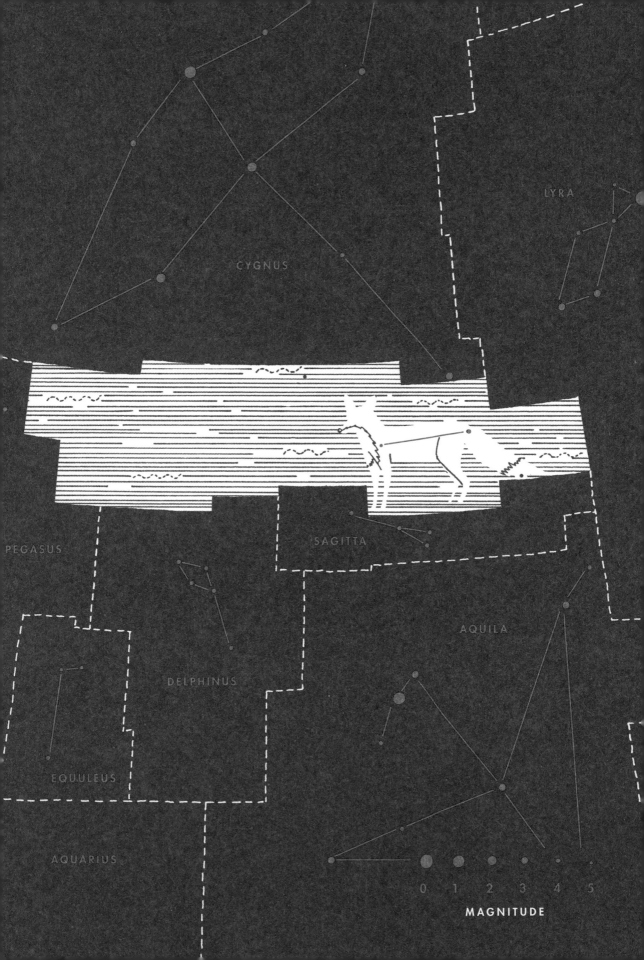

CYGNUS

LYRA

PEGASUS

SAGITTA

DELPHINUS

AQUILA

EQUULEUS

AQUARIUS

0 1 2 3 4 5

MAGNITUDE

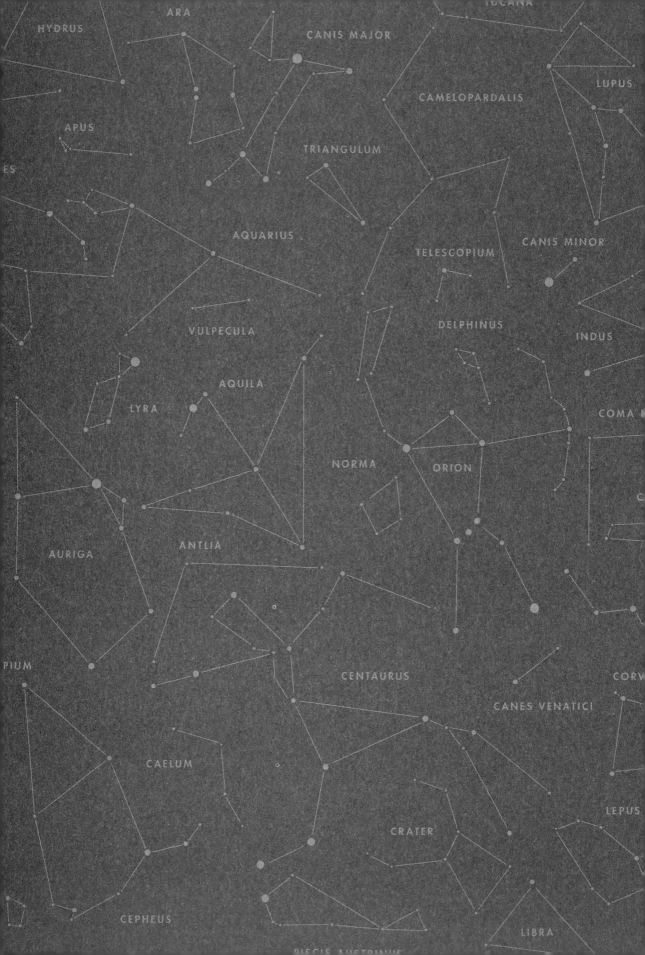

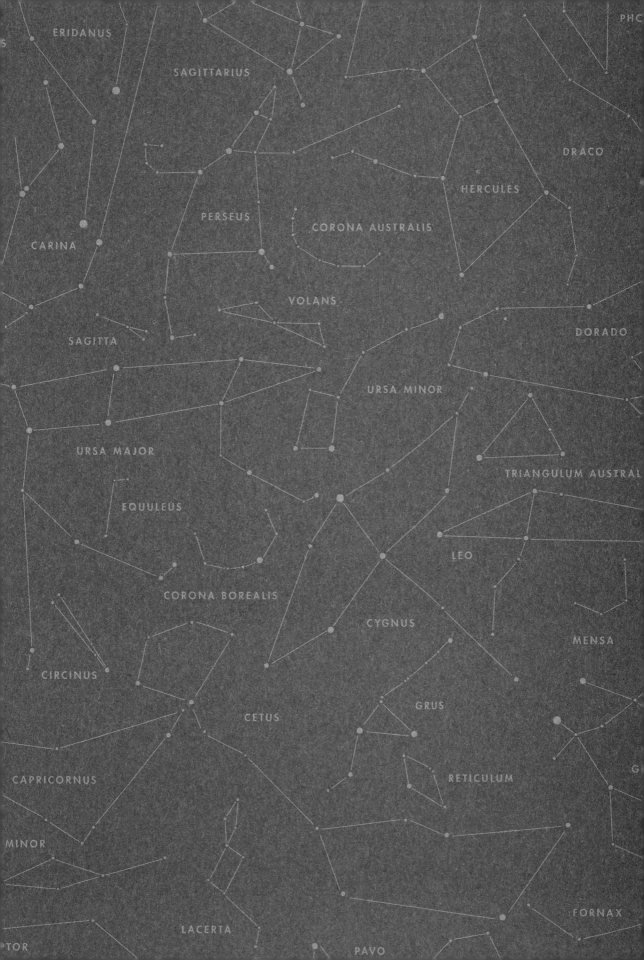

ACKNOWLEDGMENTS

WITH HEARTFELT GRATITUDE TO THE FOLLOWING PEOPLE WITHOUT WHOM THIS BOOK WOULD NOT BE WRITTEN: **FLORA ALEXANDER**, **PATRICK ALEXANDER**, **SARA BIELECKI**, **AARON DEEMER**, **KATE DONMALL**, **SOPHIE ELMHIRST**, **CHARLOTTE FAIRCLOTH**, **JANE FINNIGAN**, **CLAIRE HARRIS**, **JENNIFER HEWSON**, **ALEXANDER HISLOP**, **IAN HISLOP**, **ARTHUR HOUSE**, **SAM KNIGHT**, **ANNA LEDWICH**, **VICTORIA LEPPER**, **SARAH LUTYENS**, **THOMAS MARKS**, **MINETTE MARRIN**, **ATHENA MATHEOU**, **JULIAN MILLS**, **SUZANNE MURRAY**, **JAMES PURDON**, **BENJAMIN RAMM**, **NAOMI REYNOLDS**, **MARK RICHARDS**, **IAN RIDPATH**, **SARAH RIGBY**, **FELICITY RUBENSTEIN**, **CHARLOTTE SACHER**, **PETE SCOTT**, **ANNA STEADMAN**, **DAN STEVENS**, **EMILY STOKES**, **POLLY STOKES**, **HANNAH WALDRON**, **WILL WEBB**, **HANNAH WESTLAND**, **MAX WHITBY,** AND **CAROLINE WILLIAMS**. MY THANKS TO EVERYONE AT **HUTCHINSON**. AND WITH LOVE, ALWAYS, TO **PATRICK**.
S.H.

I WOULD LIKE TO THANK **SARAH RIGBY** AT HUTCHINSON, **SUSANNA HISLOP,** AND DESIGNER **WILL WEBB** FOR BEING SO FANTASTIC TO WORK WITH. MY THANKS ALSO TO **HUGH FROST** FOR HIS SUPPORT THROUGHOUT THE PROJECT AND FOR HIS WILLINGNESS TO MODEL ALL MANNER OF HEROIC POSES.
H.W.

THE PUBLISHERS WOULD LIKE TO THANK **SUSANNA HISLOP**, **HANNAH WALDRON,** AND **WILL WEBB** FOR THEIR UNSTINTING CREATIVITY AND ENTHUSIASM, AND **IAN RIDPATH** FOR HIS EXPERT EYE.

INDEX

47 Tucanae 180
12625 Koopman 104

A

Aboriginal Australians 60, 68, 118
Achilles 70, 78
Acrux (Alpha Crucis) 68
Admete 88
Aeneas 130, 136
Aeolus 78
Agenor, King 172
Aietes, King 154, 188
Air Pump, The 6
Akkadians 110
Alaskan Inuits 32
Alfonso X of Castile, King 158
Algenib (Gamma Pegasi) 138
Alice's Adventures in Wonderland xi, 6, 120
Allen, R. H. 178
Almagest x–xi, 18,118
Alpha Andromedae 138
Alpha Arietis 18
Alpha Böotis 22, 190–1
Alpha Canis Majoris 22, 32, 34
Alpha Canum Venaticorum 30
Alpha Columbae 56
Alpha Coronae Borealis 62
Alpha Crucis 68
Alpha Leporis 106
Alpha Orionis 134
Alpha Scorpii 164
Alpha Ursae Minoris 182, 186 *see also* North Star; Polaris; Pole Star
Alpha Virginis 190–1
Alpha Vulpeculae 196
Alphecca 62
Alpheratz (Alpha Andromedae) 138
Altair (Alpha Aquilae) x
Altar, The 16
Amaltheia the goat 20
Amazon Star (Gamma Orionis) 134
Amphitrite 72
amphora 66

anchor 68
Andromeda **2–3**, 40, 46, 49, 52, 138, 142, 158
Anser (Alpha Vulpeculae) 196
Antares (Alpha Scorpii) 164
Antlia **6**
Anubis 32
Aphrodite 58
Apis 124
Apollo 42, 64, 66, 130
Apus **8**
Aquarius 6, **10–11**, 160
Aquila 10, **14**, 43, 160
Ara **16**, 110
Arabs xi, 110, 134
Arago, François 104
Aragon, Louis 56
Arc, The 182
Arcas 110, 183, 186
Archer, The 162
Arcturus (Alpha Böotis) 22, 190–1
Ares 164, 188 *see also* Mars
Argo Navis 18, 38, 56, 154, 188
Argus 136, 172
Ariadne's Crown 62
Aries **18**
Arrow, The 160
Artemis 78, 130, 182
Ascalon 49
Asclepius 42, 43, 78, 106, 130–1, 160
Ashwins, The 18
Asses and the Manger, The 28
asterisms xii
astrology 108
Atargatis 152
Athamas, King 188
Athena 20, 76
Athene 93, 139, 142, 160, 188
Auriga **20**
Aurvandil 134
Australians 60, 68, 118
Azhdeha 76

B

Babylonians 10, 110, 190
Bacchus 34, 60 *see also* Dionysus
Bailly, Jean Sylvain 126
Bakairi Indians 164
barber 58
Bartsch, Jacob 26
Baseball Diamond, The 2, 138
Bayer, Johann 86, 98, 124, 146,
   178
Bear Cub, The 186
bees 124
beetles 28
Beggar's Dish 62
Bellatrix (Gamma Orionis) 134
Bellerophon 138
Belt, The 134
Berenice's Hair 58
Berger, John xi
Beta Columbae 56
Beta Coronae Borealis 62
Beta Crucis 68
Beta Leporis 106
Beta Pegasi 138
Beta Pictoris 146
Betelgeuse (Alpha Orionis) 134
Bier, The 182
Big Dipper, The 182, 191
binary stars 32 *see also* double stars
Bird of Paradise, The 8
birds 144, 160, 180 *see also* Bird of Paradise,
   The; Crane, The; Crow, The; Dove, The;
   Swan, The
Bissell-Thomas, James D. 6
Blaeu, Willem Janszoon 124
Board of Longitude 90
Bode, Johann 82, 146,
   166
boomerang 60
Boorong Malleefowl 52
Boötes 22, 62, 191
Botswana 68
Brahe, Tycho 54
Brazil 68, 164, 180
Browning, Sir Frederick "Boy" 138
bucket 66
Bull, The 172
Bull of Poniatowski 130
Butterfly, The 134

C

Caelum 24
Caer Arianrhod 62
Callisto 136, 182, 183, 186
Camelopardalis 26
Cancer 6, **28**, 93
Canes Venatici 30, 100
Canis Major 32, 134
Canis Minor 22, **34**, 134
Canoe of Tamarereti 134
Canum Venaticorum 30
Capricornus 36, 149
Carina 38
Carlyle, Thomas 134
Carroll, Lewis 6
cartography xi–xiii, 26, 52, 56, 118, 124,
   146
Cassiopeia 2, **40**, 46
Castor 38, 78, 84, 112, 131
cats 112
cave paintings 118
Cayman Islands 72
Ce 106
Celeris 78
Celestial Jackal 32
Celestial Prison 62
Celestial Sphere xiii, 186
centaurs 42, 162
Centaurus 42–3, 68, 78, 106, 110
Cepheus 46
Cerberus 32, 88
Ceres 176
Ceryneian Hind 88
Cetus 2, 40, 46, **48–9**, 49
Chamaeleon 52
Charioteer, The 20
China xi, 8, 32, 52, 84, 100, 106, 134, 156
Chiron 14, 43, 78, 88, 106, 130, 188
Christianity 68
Chukchi 62
Churchill, Winston 49
Circinus 54, 156
Circlet, The 148
Clark, Alvan G. 32
Coalsack Nebula 68
Cock, Mr., Philomathemat 10
Colchis 18, 36, 188
Columba 56, 106
Coma Berenices 58
compasses 54, 156

Conon of Samos 58
constellations *see* official constellations
Copernicus, Nicolaus 54, 174
Cor Scorpii (Alpha Scorpii) 164
Cornucopia 20
Corona Australis 60, 162
Corona Borealis x, **62**
Coronis 130
Corsali, Andrea 68
Corumphiza 108
Corvus 6, **64**, 66, 106
coyote 52
Crab, The 28
Crane, The 86
Crater 64, **66**
Cronos/Cronus 20, 43, 72, 162
Crosby, Stills & Nash 68
Cross, The (Southern) 68
crossword 68
Crotus 162
Crow, The 64
Crux 68
Cup, The 66
Cupid 149 *see also* Eros
Cyclopes 160
Cycnus 70, 80
Cygnus 70, 84

D

de Houtman, Frederick 8, 52, 74, 86, 98, 124, 136, 144, 146, 178, 180
Delphinus 72, 78
Delporte, Eugène xii, 118
Delta Capricorni 36
Delta Coronae Borealis 62
Delta Crucis 68
Delta Leporis 106, 176
Deltoton 176
Demeter 191
Deneb Algedi (Delta Capricorni) x, 36
Devil, the 116
Diamond (of Virgo), The 30, 102, 190
Diana 88
Diomedes, King 88, 188
Dionysus 62 *see also* Bacchus
Dithutlwa (Crux) 68
Djobela 116
Dog Days 32

Dog Star (Alpha Canis Majoris) 32
dogs 30, 34, 110 *see also* Cerberus
Dolphin, The 72
Donaghy, Michael 174
Donne, John 54
Dorado 74
double stars 18 *see also* binary stars
Dove, The 56
Draco 76, 80
Dragon, The 76
Druids 182
du Maurier, Daphne 138
Dubhe 182
ducks 68
dung beetles 28
dwarf galaxies 180

E

Eagle, The 14
easel and palette 146
Egyptians xi, 10, 58, 108, 128, 164, 190
emu 68
Enif (Epsilon Pegasi) 138
Enki 16
Epsilon Coronae Borealis 62
Epsilon Leporis 106
Epsilon Pegasi 138
Equuleus 78
Equuleus Pictorius 146
Equus Primus 78
Eratosthenes 176
Erichthonius 20
Eridanus 70, **80**
Eros 160 *see also* Cupid
Ethiopia 80
Ethiopian King, The 46
Europa 172
Eurystheus, King 88
Evil One, the 76

F

False Cross, The 38
Family, The 14
feces 106
Field, Katie 60
Firmamentum Sobiescianum 104

First Star of Aries (Gamma Arietis) 18
fish 62, 148, 152, 194 *see also* Flying Fish, The;
  Goldfish, The
Fish Hook, The 164
Fishes, The 148
fishing rod 86
flags 68
flamingo 86
Flammarion, Camille 146
Flamsteed, John 146
Fly, The 124
Flying Fish, The 194
Fornax 82
fountain 139
foxes 32
Frederik's Glory 2
Freemasonry 126
French Revolution, The 82
Frying Pan 52
Furnace, The 82

G

Gacrux (Gamma Crucis) 68
Gaia 20, 80, 149
Galileo 54, 174
Gamma Arietis 18
Gamma Coronae Borealis 62
Gamma Crucis 68
Gamma Leporis 106
Gamma Pegasi 138
Ganymede 10–11, 14, 160
geese 196
Gemini 6, 78, **84**
Geryon 88
Gilgamesh 134
Giraffe, The 26
Glaucus 131
goats 20, 36, 162
Godfrey, Thomas 128
Golden Fleece 18, 38, 84, 188
Goldfish, The 74
Great Bear 80, 182
Great Square, The 2, 138
Greater Dog, The 32
Greeks x–xii, 8, 10, 11, 18, 22, 32, 60, 66, 68,
  72, 84, 108, 110, 112, 136, 152, 160, 162,
  176, 186, 188
Greenwich Mean Time 90

Grus **86**
Guardians of the Pole, The 186
Guernica 56

H

Hades 72, 131, 191
Hadley, John 128
haiku 194
Halley, Edmond 128
Hamal (Alpha Arietis) 18
Hare, The 106
Harrison, John 90
Head, The 48, 76, 92
Heavenly G, The 20, 32, 34, 84, 134, 172
Hebe 10
Helios 80, 191
Helle 18, 188
Henrietta 30
Hephaestus 20, 22, 62, 160
Hera 10, 11, 28, 136, 172, 182, 183, 186 *see*
  also Juno
Hercules 14, 43, 62, 76, **88**, 93, 102, 110, 134
  as Heracles 38, 154, 160
  as Herakles 28
Herdsman, The 22
Hermes 78, 136, 142, 172 *see also* Mercury
Hero, The 142
Herschel, John 166
Hesperides 88
Hevelius, Johannes 30, 100, 104, 112, 168,
  170, 196
  fire at his home 168, 170
Hieropolis 152
Hipparchus x, 100
Hippe 78
Hippocrene 139
Hippolyte 88
Hippolytus 130
Hiroshima 86
Hokule'a (Alpha Böotis) 22
Homam (Zeta Pegasi) 138
Homer 22
Hooke, Robert 18
horns 20
Horologium 90, 146
horoscopes 108
Horse and Rider, The 182
horses 78, 88

Hughes, Ted 70
Hunter, The 134
hunting 106, 134
Hunting Dogs, The 30
husbands 64
Huygens, Christiaan 90
Hyades, The 118, 172
Hydra 6, 64, 66, 88, **92–3**, 93
Hydrus **96**, 128, 146
Hygieia 42
Hyginus 22, 176

I

Icarius 22, 34
Idas 112
Igbo-Ora 84
Indian, The 98
indigenous peoples 60, 98, 118, 134, 164, 178
Indonesia 180
Indus **98**
International Astronomical Union xii, 118
Inuits 32, 134
Io 136, 172
Iolaus 93
Iota Corona Borealis 62
Ischys 130
Ishtar 190 see also Virgo
Isles of Scilly 90

J

Japan 86
Jason and the Argonauts 18, 38, 56, 84, 154, 188
Jeremiah VIII 86
Job's Coffin 72
jokes 93
Joppa, king of 46
Juno 60, 136 see also Hera
Jupiter 60, 136

K

Kahnweiler, Daniel-Henry 56
Keats, John 52, 139
Keel, The 38

Kepler, Johannes 54, 174
Keyser, Pieter Dirkszoon 8, 52, 74, 86, 98, 124, 136, 144, 146, 178, 180, 194
Kids, The 20
Kneeling Hero, The 88
Koopman, Catherina Elisabetha 104, 168, 170
krater 66

L

Lacaille, Nicolas Louis de 6, 24, 38, 54, 82, 90, 116, 118, 126, 128, 146, 154, 156, 158, 166, 174
Lacerta **100**
Ladon 76, 88
Lalande, Jérôme 126
Lambda Sagittarii 162
Large Dipper, The 2, 138, 142
Large Magellanic Cloud 74, 116
Larkin, Philip 174
Lavoisier, Antoine 82, 126
Le Verrier, Urbain 36
Leda 70, 84
Leo 6, **102**
Leo Minor 100, **104**
Lepus 6, 32, **106**
Lesser Dog, The 34
Lesser Water-Snake, The 96, 128
Libra **108**
Libya 80
limericks 40
Lion, The 102
Lion Cub, The 104
Lipperhey, Hans 174
Little Bear 80
Little Dipper 52, 186
Little Fox, The 196
Little Horse, The 78
Lizard, The 100
lodestone 156
longitude 90, 108, 178
Lozenge, The 76
lucida definition xiii
Lupus **110**
Lycaon, King 110, 183
Lynceus 112
Lynx 100, **112**
Lyra **114**, 156
Lyre, The 114

M

M31 2–3
Mackiernan, Douglas S. 128
Madagascar 98
Maera 22, 32, 34
Magellan, Ferdinand 8, 180
Magic Flute, The 126
magnetism 156
magnitude xii
Maiden, The 190
Maori 68, 84, 134, 164
Marduk 48, 76
marine chronometer 90
Markab (Alpha Pegasi) 138
Mars 164 *see also* Ares
Marshall Islands 86
Matisse, Henri 56
Maui 164
Medea 38, 154
medicine, symbol of 131
Medusa 46, 139, 142
Melanippe 78
Menkib (Beta Pegasi) 138
Mensa 116, 158
Merak 182
Mercury 136 *see also* Hermes
mermaids 152
Mesarthim (Gamma Arietis) 18
Mesopotamia 16
Meteorologiae 10
Microscope, The 118
Microscopium 118, 158
Milk Dipper, The 162
Milky Way, The 14, 16, 68, 74, 162
Milton, John 130
Mimosa (Beta Crucis) 68
Minos, King 88, 131
Minotaur 88
Monoceros 6, **120–1**
Montaigne, Michel de 98
Moon Dog 32
mother and baby 164
Mount Atlas 88
Mount Helicon 139, 162, 188
Mount Ida 20
Mount Olympus 14, 60, 110, 142
Mozart, W. A. 126
Mu Leporis 106
Musca 124
muses 162

Myrtilus 20
myrtle 60

N

Native Americans 182
Navajo 52, 100
navigation xi, 28, 90, 128, 156, 178
Nazi concentration camp 84
Neptune 36, 70, 88 *see also* Poseidon
Neptuni Proles 36
Nereids 46, 72
Net, The 158
New Zealand 68, 134, 164 *see also* Maori
newt 100
Nigeria 84
Norma 54, **126**
North Star 182 *see also* Alpha Ursae Minoris
Northern Cross, The 70
Northern Crown, The 62
Northern Fly, The 18
northern hemisphere xiii, 2, 68, 164, 182, 186

O

O'Casey, Sean 136
Octans 128, 146
Octant, The 128
Odysseus 22
official constellations xii, 118
opera 126
Ophiuchus 66, **130–1**, 160
origami 86
Orion xi, 32, 34, 106, **134**, 164
Orpheus 114, 154
Ovid 3, 14, 52, 60, 136, 144

P

paganism 120–1, 190
Painter's Easel, The 146
Pan 36, 149, 162
Papin, Denis 6
Papua New Guinea 8, 68
Parker, K. Langloh 60
Pascal, Blaise 54
Pavo 136, 158

peace, symbol of 56
Peacock, The 136
Pegasus 46, 78, **138–9**, 142
Pelias 188
Pendulum Clock, The 90
Peregrinus, Peter 156
Persephone 190–1
Perseus 3, 40, 46, 49, 139, **142**
Persian 66, 108
Phaethon 70, 80
Phakt (Alpha Columbae) 56
Phi Sagittarii 162
Philippines 180
Philyra 43
Phineus 154, 162
Phoenicians 186
Phoenix 8, **144**
Phrixus 18, 188
Picasso, Pablo 56
Pictor 146
Pigafetta, Antonia 8
Pisces 2, **148–9**
Piscis Austrinus 152
Plancius, Petrus 26, 56, 86, 98, 124, 178, 180
planetary configurations 108
Planisphere 82, 146, 166, 178
Plath, Sylvia 174
Pleiades, The 118, 172
Plow xii, 22, 182
Pluteum Pictoris 146
Pluto 176
poetry 54, 58, 70, 130, 139, 174
Pointers, The 182
polar bear's paw 62
Polaris (Alpha Ursae Minoris) 182, 186 *see also* Pole Star
Pole Star (Alpha Ursae Minoris) 76, 182, 186 *see also* Polaris
Pollux 78, 84
Polydectes, King 142
Polydeuces 38, 112, 154
Polynesian 22, 178
Pope, Alexander 58
Portuguese 74
Poseidon 36, 40, 46, 72, 78, 139, 186 *see also* Neptune
position of the stars, moving 186
precession 186
Procyon (Alpha Canis Majoris) 34
Prometheus 14, 43, 78, 160

Proserpina 176
Ptolemy x–xi, 18, 38, 66, 68, 110, 118, 134, 146, 154, 186
Pullman, Philip 156
Pup 32
Puppis 154
Pythagoras 144
Pyxis 156

Q

Qamata 116

R

R136a1:10 74
Rake, The 134
ram 188
Ram, The 18
Rebekah 26
Rehua (Alpha Scorpii) 164
Reticulum 146, **158**
Rhea 20, 43
Ridpath, Ian 178
Rigel 134
River, The 80
river-crossing puzzle 196
Roger of Hoveden 108
Romans xi, 32, 36, 110, 144, 160, 176
Royal Society, the 126
Rubens, Peter Paul 3

S

Sagitta 160
Sagittarius x, **162**
Sail, The 64, 188
St. Paul's Cathedral 54
Samoa 68
Sasaki, Sadako 86
satellite galaxies 74
Saucepan 182
Scales, The 108
Scheat (Beta Pegasi) 138
scientific instruments 118, 126, 128, 158, 170, 178
Scorpio 80

Scorpion, The 164
Scorpius x, 106, **164**
script 92
Sculptor **166**
Sculptor's Chisel, The 24
Scutum 100, **168**
Sea Goat, The 36
sea of stars 10
Seago, Edward 138
Sea Monster, The 48
Segment, The 142
Semele 60
Serpens **130–1**
Serpent, The 76
Serpent Bearer, The 130–1
serpents 100, 131
Set Square, The 126
Sextans 100, **170**
sextant 22, 112, 128, 170
Shakespeare, William 98
Shen 134
Sheratan 18
Shield, The 168
Shipping Forecast, the 90
Shuruppak 16
Siberian 62
Sicilia 176
Sicily 176, 190
Sickle, The 102
Sidney, Philip 174
Sigma Sagittarii 162
Sikuliaqsiujuittuq (Alpha Canis Majoris) 34
Sirius (Alpha Canis Majoris) 22, 32
Sirrah (Alpha Andromedae) 138
skinker 10
Small Magellanic Cloud 180
Smith, Patti 68
snakes *see* serpents
Sobel, Dava 90
Sobieski III, King John 168
Sopdet (Alpha Canis Majoris) 32
Sophocles 22
South Africa 82, 116
Southern Cross, The 68
Southern Crown, The 60
Southern Fish, The 152
southern hemisphere xiii, 68, 124, 136, 178, 182
Southern Triangle, The 178
Spica (Alpha Virginis) 190
spiral galaxies 2, 3

Spring Triangle, The 102, 190
star atlases xi–xii, 3, 98, 100, 104, 124, 146, 168, 176, 178
stargazer's adage 190
statues 72
Stern, The 154
stork 86
storms 16, 108
Sula (Crux) 68
Sumerians 164
Summer Triangle, The x, 14, 70, 114
Sunset Reef 72
supergiants 134, 138, 164
Swan, The 70
Sword, The 134
Syrian refugee camp 152
Syrinx 36

T

Table Mountain, The 116
Tahiti 84
Tarantula Nebula 74
Tau Sagittarii 162
Taurus 106, 134, 136, **172**
Te Punga (Crux) 68
teapot 162
telegram 142
telescopes 32, 112, 146, 158, 168, 174
Telescopium **174**
Telesphorus 42
Tengshe 100
Tennyson, Alfred Lord 32
Thales 76
Theseus 62
Theta Coronae Borealis 62
Thixo 116
Thompson, D'Arcy 106
Thor 134
Three Guides, The 2, 40
Three Kings, The 134
Three Patriarchs, The 178
Through the Looking-Glass 6, 121
Tiamat 49, 76
Tianxiang 170
T'ien-kaou 32
T'ien-lang 32
T'ien-lao 62
T'ien-wei 62

Toloa (Crux) 68
Tongans 68
tortoise 52
Toucan, The 74, 180
Triangle, The 176
Triangulum **176**
Triangulum Australe **178**
Trinacria 176
Tucana 6, **180**
Tūhoe 164
Tupi 180
Turkish 110
Twins, The 84
Tyndareus 131
Typhon 36, 149

U

Unicorn, The 120
Urania 100
Uranometria 98, 124, 178
Uranus 20
Ursa Major xi, xii, 22, 136, **182–3**, 191
Ursa Minor 22, 76, **186**
Utnapishtim, King 16

V

V, The 172
van den Keere, Pieter 86
van Hunks, Jan 116
Vega (Alpha Lyrae) x
Vela **188**
Venus 149
Venus's Mirror 134
Vespucci, Amerigo 178
Vikings 128
Virgil 32, 136
Virgo 6, 22, 176, **190–1**

Volans **194**
Vulpecula 100, 168, **196**

W

Wasn (Beta Columbae) 56
Water-Bearer (or Jar), The 10
Water-Snake, The 92
Welsh 62
whale 48
Wilde, Oscar 52
wine cup 66
Winged Horse, The 138
Winter Octagon, The 20, 32, 34, 84,
   134, 172
Winter Oval, The 20, 32, 34, 84, 134, 172
Winter Triangle, The 32, 34, 134
Wolf, The 110
World Peace Congress 56

X

Xhosa 116

Y

Y, The 190
Yeats, W. B. 70
Yin and Yang 84

Z

Zeta Pegasi 138
Zeta Sagittarii 162
Zeus 10–11, 14, 18, 20, 22, 36, 43, 66, 70,
   72, 80, 84, 110, 112, 130, 160, 162, 172,
   182, 186, 191

ABOUT THE AUTHOR AND ILLUSTRATOR

··

SUSANNA HISLOP IS AN ACTOR, WRITER, DIRECTOR, AND THEATRE MAKER. SHE IS AN EDITOR OF THE ONLINE QUARTERLY *THE JUNKET* AND ARTISTIC DIRECTOR OF SLIP OF STEEL THEATRE COMPANY.

HANNAH WALDRON IS AN ARTIST AND DESIGNER BASED BETWEEN LONDON, UK, AND STOCKHOLM, SWEDEN. HER GRAPHIC- AND NARRATIVE-LED IMAGE MAKING HAS BEEN APPLIED ACROSS A RANGE OF MEDIA, FROM PRINT TO TEXTILES, AT BOTH A PERSONAL AND ARCHITECTURAL SCALE. SHE HAS FOUND WEAVING TO BE A NATURAL COMPLEMENT TO HER GRID-BASED IMAGE MAKING AND HAS RECENTLY COMPLETED AN MFA IN TEXTILES IN THE EXPANDED FIELD.

'LYING ON OUR BACKS,
WE LOOK UP AT THE NIGHT SKY.
THIS IS WHERE STORIES BEGAN...'

JOHN BERGER